Electrochemical Energy Storage

Energy Storage – Batteries and Supercapacitors Set

coordinated by
Patrice Simon & Jean-Marie Tarascon

Volume 1

Electrochemical Energy Storage

Jean-Marie Tarascon
Patrice Simon

iSTE

WILEY

First published 2015 in Great Britain and the United States by ISTE Ltd and John Wiley & Sons, Inc.

ISTE Ltd
27-37 St George's Road
London SW19 4EU
UK

www.iste.co.uk

John Wiley & Sons, Inc.
111 River Street
Hoboken, NJ 07030
USA

www.wiley.com

Library of Congress Control Number: 2014958638

British Library Cataloguing-in-Publication Data
A CIP record for this book is available from the British Library
ISBN 978-1-84821-720-1

Contents

INTRODUCTION vii

CHAPTER 1. BATTERIES AND SUPERCAPACITORS: SOME REMINDERS 1

1.1. Main evolution of batteries from the 1980s to now 1
1.2. Supercapacitors: recent developments 8

CHAPTER 2. ADVANCED LI-ION 11

2.1. Positive electrode materials for Li-ion technology 11
2.2. Negative electrode materials for Li-ion technology 14
2.3. The question of electrolytes for Li-ion technology 15

CHAPTER 3. CAPACITIVE STORAGE 17

3.1. Carbonated materials for capacitive storage 17
3.2. Pseudocapacitive materials 18
3.3. Electrolytes for supercapacitors 20
3.4. Hybrid systems and middle-term goals 21

CHAPTER 4. NEW CHEMISTRIES 23

4.1. Li-air technology 24

4.2. Li-S technology 27
4.3. Na-ion technology 29
4.4. Redox-flow technology 32
4.5. All-solid state batteries 36

CHAPTER 5. ECO-COMPATIBLE STORAGE 41

5.1. Ionothermal synthesis 42
5.2. Bioinspired synthesis/approach 42
5.3. Organic electrodes for "green" Li-ion batteries and more durable batteries 45
5.4. Recycling and LCA 47

CHAPTER 6. SMART MATERIALS 49

6.1. Photonics of insertion materials to create photo-rechargeable batteries 50
6.2. Micro-energy sources 51

CHAPTER 7. TECHNOLOGY TRANSFER, RESEARCH PROMOTION AND EDUCATION 53

7.1. Development: industrial property 53
7.2. Education 54
7.2.1. Erasmus Mundus Master's degree: Materials for Energy Storage and Conversion (MESC) 55
7.2.2. Specialization in Energy Storage and Conversion (SCE), at ENSCBP (Bordeaux – INP) 57

CONCLUSION 59

BIBLIOGRAPHY 63

INDEX 75

Introduction

The supply and management of energy are more than ever at the center of our daily concerns and are a major socioeconomic priority. Currently, we depend on fossil fuels with two serious consequences: exhaustion of reserves and worsening of the greenhouse effect caused by the emission of carbon dioxide (CO_2) from their combustion. Due to an increasing world population, which is expected to expand from 7 to 10 billion by 2050, and economic development, energy demands will double, from 14 TW currently to 28 TW in 2050, which would increase the atmospheric concentration of CO_2 if no action is taken; the result would be ever greater climate warming [TAR 11]. Thus, all global organizations agree that energy is the main challenge of the 21st Century that our planet must overcome.

Nowadays, if we have any hope of reversing this trend, we must develop the use of renewable energies (solar, wind, geothermal, biomass, etc.), which, despite their intermittent character, have a low CO_2 footprint. However, for this energy transition to be successful, it is important to consider how they can be used more efficiently and find innovative management solutions, reliable conversion and storage of energy, that are low cost and widely applicable. For this, we need (1) efficient photovoltaic and thermoelectrical systems to convert light and heat into electricity, respectively;

(2) electronic conductors such as superconductors to minimize the Joule effect; and (3) storage systems such as batteries/supercapacitors to store energy in chemical forms and convert it back to electricity when required. Although these issues associated with the production, transport and storage of energy are exciting due to the different strategies implemented, we will address only electrochemical storage in this book. The storage of electrical energy will continue to play an increasingly vital role in sectors such as transport (electric and hybrid vehicles), medicine, defense and aerospace, telecommunications and other sectors. In 2020, it is predicted that, for example, 10% of cars produced will be electric and 20% of the energy used worldwide will come from renewable energy. The storage and production of electrical energy are crucial elements in a completely new paradigm of energy. It has become an important and strategic issue for France and its industry, as noted during the French national debate on the energy transition and the drafting of the upcoming law.

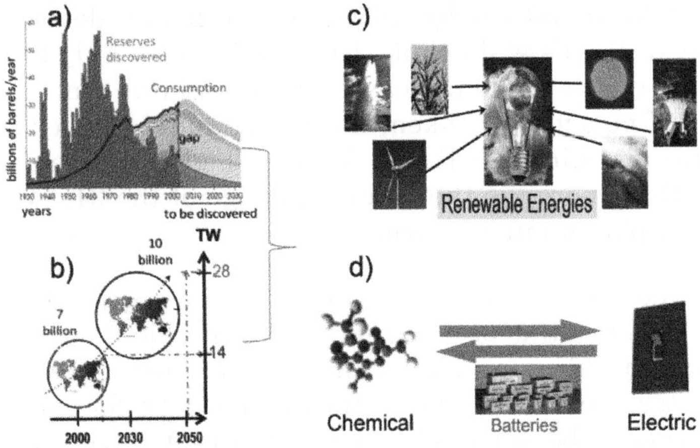

Figure I.1. *The decrease in fossil fuels a) associated with the increasing demand for energy b) makes renewable energy c) a solution for the successful energy transition as long as we can compensate for their intermittence using electrochemical devices d)*

These two applications (networks and transport) must therefore store energy and convert it back to an electrical form. One of the best ways of doing this is to convert chemical energy into electrical energy since they both share the same vector, the electron. Electrochemical devices capable of doing this conversion are known as fuel cells, supercapacitors and batteries. More specifically:

– *fuel cells* operate based on the reverse principle to the electrolysis of water, i.e. the electricity is produced by oxidation on a di-hydrogen (H_2) electrode coupled with reduction on another electrode of an oxidant such as oxygen from the air, together producing water. This is an "open" system, that is to say, directly supplied externally, and hence not directly electrically rechargeable;

– *supercapacitors* are based on capacitive properties of a double electron-ion layer at the electrolyte–electrode interfaces with a capacity per unit mass or area expressed in $F.g^{-1}$ and $F.cm^{-3}$, respectively, up to millions of times greater than that of typical capacitors;

– finally, *accumulators* [TAR 98], simplified to "batteries" by misuse of language, can deliver/store electrical energy generated from reversible redox reactions that may occur in the constituent materials of their electrodes. They are characterized by (1) an output potential (V_{out}) which is the difference in potential of the redox reactions occurring at the positive and negative electrode and (2) a mass energy density (Wh/kg) which is the product of the potential (V_{out}) per mass capacity of the electrode, expressed in mAh/g.

Based on the Ragone diagram, we can see that for power applications, first come supercapacitors, then batteries and then fuel cells, whereas in terms of autonomy, it is the opposite sequence that prevails. Another important piece of information on this diagram is that none of these systems can currently compete with internal combustion engines. This is why, in the beginning of the 20th Century, electric vehicles gave way to internal combustion vehicles, which took advantage of their low cost, ease of use and the high performance of fossil fuels (1 L of oil still provides 15 times more energy than that of Li-ion technology). There is still a long way to go before electrochemical storage can provide the energy of oil at the same mass.

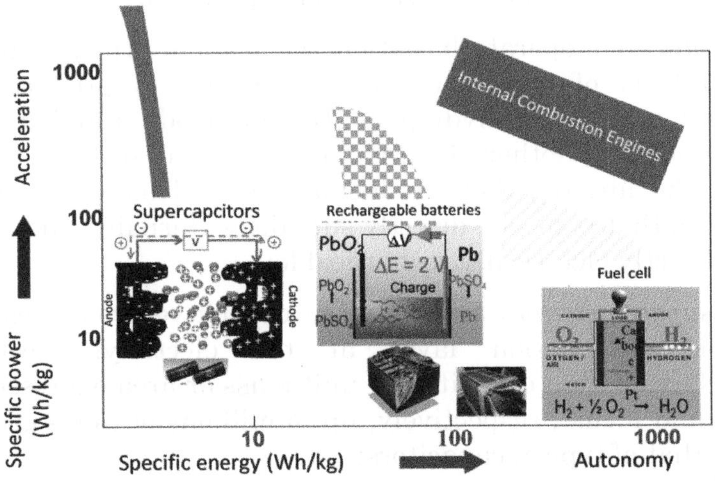

Figure I.2. *The different electrochemical devices, super capacitors, batteries and fuel cells are compared with fuel motors within the same Ragone diagram*

This is a complex task given that the progress in the field of energy storage has been slow [TAR 98]. It has been almost 200 years since the fuel cell was discovered by W. Groove in 1839, but some questions still remain unanswered. The history of batteries is better known; however, their energy density has only increased by four or even five times since the discovery of the lead battery in 1859 by Gaston Planté. A common theme in electrochemical systems is that their performance is (1) dependent on the quality of the electrode–electrolyte interfaces, which are still difficult to fully understand, and (2) dependent on the ability of the chemist to design better performing materials on request.

Therefore, technological breakthroughs are required to speed up the increasingly slow progress and even more so since the challenges posed by the increasing demand for energy consumption with low climatic impacts are pressing. Our society is now aware of this and the electrochemical storage of energy has become a major strategic issue for our nation, thereby motivating major innovative projects, one of which brought together French researchers and privately-held companies in the domain of batteries and supercapacitors, as described below.

Creation of the French network on electrochemical energy storage (RS2E)

France has considerable resources to meet the challenges in the domain of electrochemical energy storage, most notably in the field of electrode materials for batteries and supercapacitors, which were developed by the French National Center for Scientific Research (CNRS) and universities. Public organizations also need to be mentioned, such as the French atomic energy and alternative energies commission (CEA), which can technologically use these findings to create innovative systems, and a wide range of industries, whether in the field of transport, energy production, material chemistry or electronics, all of which may benefit from improved synergy.

However, a quick look at our past reveals that our scientific advances regarding the materials of electrodes ($LiCoO_2$, $LiFePO_4$, etc.) as well as electrolytes (salts and polymers) have become commercial products outside France, especially in Asia, which currently dominates the accumulator market. To change the current situation and reverse the course of history, the French government has tried to strongly develop the innovation potential in the field of energy storage, through a coordinated intensification of research, development and industrialization, and also by bringing together academic laboratories and technological research centers. Thus, on the 2nd of July 2010, the French network on electrochemical energy storage (RS2E) was created. It is based on an integrated vision (Figure I.3) combining research excellence and innovation of national research labs (17 CNRS/ Universities joint-laboratories) together with efficient and experienced technological research centers (CEA, IFP[1] and INERIS[2]), all being connected to a reinforced industrial network. Its main aim is based on the motto "integrate, unite, and gather to better innovate and develop", in order to promote the emergence of a new generation of systems with the desire to see France become a global leader in a field as economically lucrative as that of energy.

Figure I.3. *Integrated vision of RS2E with left: the different laboratories whose mission is to explore new concepts of batteries and supercapacitors and more specifically high-performing materials, thereby creating new knowledge and stimulating numerous scientific innovations, middle: technological research public organizations which aim to develop, on a large scale, the exploration of functional prototypes and demonstrate innovative systems in order to transfer technological breakthroughs into the industry and, right: Industrial Partners, with the logos of the different members as of September 2014*

1 IFP: French institute of petroleum and new energies.

2 INERIS: National competence centre for industrial safety and environmental protection.

This collaborative network has allowed the development of a structure, an environment and, in particular, a feasible strategy:

– to cultivate and expand the platform of scientific and technological knowledge in the field of electrochemical storage of energy;

– to unite experts, including foreigners, with a variety of cultures and backgrounds to promote creativity and innovation and to form a center of expertise at the national and international level directly linked to industry;

– to use these different cultures for (1) further research and basic research breakthroughs, (2) the development of new concepts, (3) the design of materials with enhanced properties, (4) cultivate a visionary approach suitable for the development of technologies beyond lithium, overall forming a center of expertise for our industries to carry out generic research but also better adapted to their specific needs;

– to propose, via a game of continuous "Ping-Pong" between the different partners, a rapid and efficient transfer of new materials and developed concepts.

Strategy and scientific structure of RS2E

Apart from an integrated vision that provides the basis and connectivity needed between the different stakeholders from the academic, technological and industrial sectors, RS2E is also based on an integrated scientific program, for the development of high-performing energy storage systems, aiming:

– to improve the existing materials, research of new materials/redox systems and the development of new concepts. The success in this field will depend on our capacity not only to understand the reaction mechanisms at play but also to integrate the component size and system into our research;

– to encourage risk-taking through the launch of new topics to motivate technological breakthroughs, which could be the flagship of global and French research in the coming decades.

A more efficient device is synonymous with an affordable high energy density system. It is clear that any innovation in this domain must take into consideration the production cost of these systems, which depends on the abundance of materials and production processes. Investigating *other chemicals apart from Li* or *double-layer supercapacitors* is vital to obtain energy density or exacerbated capacitance systems, respectively. Increasing the *sustainability and lifespan of devices* is also essential for their users. The same goes for *recycling issues* with regard to sustainable development. Such requirements add to the limitations while providing new opportunities as well as *new concepts such as renewable organic electrodes*, or even *multifunctional electrodes* uniting the storage and conversion functions in the same material. The time constraint in which we operate (with 2050 as the target date) also calls for (1) the development of reasoned *theoretical combinatorial chemistry* to rapidly identify promising electrode materials, (2) the development of *in situ characterization techniques* to understand/control interfaces. Although non-exhaustive, these are the essential criteria to make materials useful, which, through the development of a *science for the formulation of electrodes* and *efficient transfer methods*, will allow the rapid emergence of specialized national production capabilities of batteries or supercapacitors. These aspects are included in our five main research areas (Figure I.4) which will be discussed later.

The development of a device is a complex process and requires both (1) a multidisciplinary approach with use of experts in material science, electrochemists, organic chemists, physicists and electronics engineers and (2) a comprehensive approach that no partners can achieve alone. It is in the light of all these questions, and a joint analysis discussed with technological and industrial partners, that we have established the RS2E scientific strategy (Figure I.4).

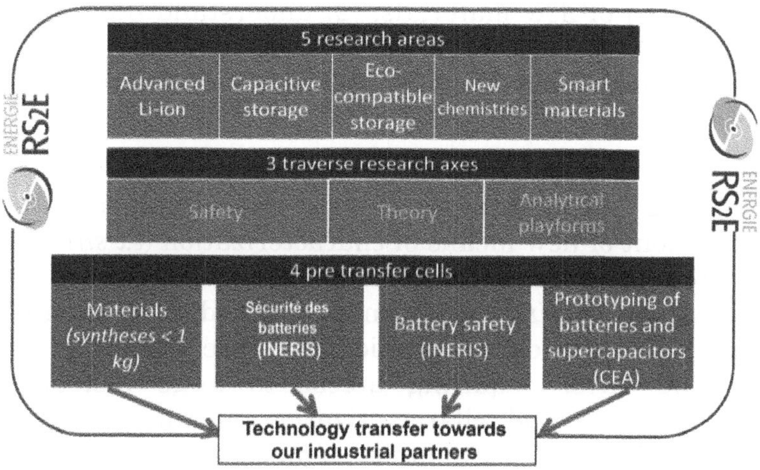

Figure I.4. *Integrated research within RS2E which is based around five areas of research (top) supported by three transverse axes (middle) and four pre-transfer units (bottom) which act as a relay with the three technological research organizations and is open to all Industrial Partners*

The organization of the network's scientific research therefore includes *3 transverse axes* at its center which form the basis of RS2E by playing the role of integration vectors of different teams while connecting our research upstream and downstream. Their missions are:

– *safety:* the activities of this transverse axis are mainly centered around (1) establishing reliable electrochemical test protocols to predict their electrochemical and thermal reliability, (2) postmortem studies of supercapacitors and batteries to determine the reasons for their failure and to provide new solutions through the addition of additives or formation tests;

– *theory:* however important the chemical intuition may be, it still finds its sources from the work of theorists. Listening to experimentalists about the problems of energy storage, this "theory" group aims to (1) define the trends in the research of new materials, (2) understand the reactivity of surfaces, (3) model the interfaces and (4) predict the electrochemical and chemical stability of electrolytes and others;

– *analytical platforms:* just as important as developing interesting new materials with exacerbated properties is the development of new means of characterization (*ex situ* and *in situ*) from a technological standpoint. Among the different methods are *in situ* development methods: (1) imaging of batteries via nuclear magnetic resonance (NMR) or X-ray (XR), the latter requiring access to large instruments, (2) observation and characterization of the evolution of interfaces (morphology and composition) using high-resolution transmission electron microscopy (HRTEM) or high resolution electron energy loss spectroscopy (HREELS), among others.

These transverse axes include constant interactions with pre-transfer "cells" which are *unique* to RS2E; their aim is to ensure the rapid transfer of a new material, electrolyte, or even a new concept to its validation by a demonstrator (prototype). In order to do this, they unite the basic links needed for a vital chain process to ensure the validity of a breakthrough. These platforms allow researchers, engineers and industries from different backgrounds to exchange, collaborate and work on the same site, with light equipment available for the prequalification of technologies before a possible transfer. These platforms are destined to become the showcase of RS2E activities to the industry. There will be four of them as follows:

– *Preprototyping of batteries and supercapacitors:* this aims at integrating innovations performed in basic research with regard to electrodes and electrolytes as well as their design on 18650-type battery elements, C-type supercapacitors, "pouch cells" and other format (Bellcore plastic technology) to obtain the reliable results required to predict the pre-industrialization upscaling.

– *Safety of prototype batteries:* it tries to integrate risk management throughout the lifecycle of new materials (maintenance, safety, toxicity and recycling) at the early stages of integration.

– *Behavioral modeling of preprototypes:* its purpose is to optimize the energy management used (electric vehicles, buses) but also diagnose the state of charge and health of accumulators while taking into account aging, overloading, deep discharge, fast charging process etc.

– *Pre-transfer materials (syntheses < 1 kg):* this is the point of entry for any industry wishing to understand the possibility of large-scale manufacturing of new materials, new electrolytes, new salts or new patented polymers in the network.

Finally, these transverse axes serve as centers of expertise in material science, *in situ* characterization tools and theoretical analyses to supply a downstream research program made up of five research topics with well-defined objectives. These *topics* have been defined based on scientific criteria aiming to (1) unite certain current activities to give them worldwide exposure, (2) allow subjects to emerge that were previously abandoned due to the underestimation of their potential or lack of critical mass to have an impact and (3) launch new topics based on our vision of the future resulting from numerous discussions between researchers and industrials.

This scientific program was established only on the knowledge of the state of the art with regard to both batteries and supercapacitors, which we will address later.

1

Batteries and Supercapacitors: Some Reminders

1.1. Main evolution of batteries from the 1980s to now

The concept and commercialization of a lithium-ion (Li-ion) battery dates back to 1980 and 1990, respectively. Its specificity resides in the fact that it uses: (1) a non-aqueous electrolyte and (2) positive and negative electrodes insertion compounds, which can act like lithium "sponges" [TAR 98]. A historical overview of accumulators shows us that this technology emerged through the continuous evolution of concepts that previously led to Pb, Ni-Cd, Ni-MeH, Li-metal and Li-polymer technologies. The lead–acid accumulator, which uses an aqueous electrolyte based on H_2SO_4 consumed during discharge, is based on conversion reactions. These reactions cause the breakage and formation of bonds, $PbO_2 \rightarrow Pb(SO_4)$ and $Pb \rightarrow PbSO_4$, at the positive and negative electrode, respectively. The transition to Ni-Cd accumulators has not only facilitated the transition from an electrolyte saturated in potassium that does not participate in the electrochemistry of a system, but also introduced a $Ni(OH)_2$ positive electrode, which uses the deinsertion and insertion of protons during charge and discharge, respectively. Insertion–deinsertion electrodes then made their first appearance. This concept was generalized some years later (1980) when Cd was replaced at the negative electrode of the Ni-Cd accumulator, by a $LaNi_5H_x$ intermetallic compound, capable of reversibly inserting–deinserting 5 H^+. Then the NiMeH accumulator emerged, the most efficient at the time (~80 Wh/kg, ~350Wh/l), which is nothing other than a proton accumulator.

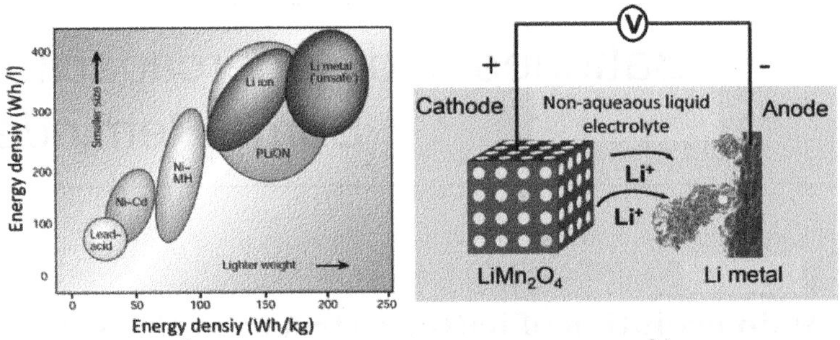

a) b)

Figure 1.1. *Comparison of different accumulators in terms of mass energy (Wh / kg) and volumic energy a) with in b) a description of the Li-metal accumulator. The dendrites are shown*

At the same time, parallel research on solid-state chemistry revealed compounds that were capable of reversibly inserting Li^+, such as Li_xTiS_2. With Li being one of the most electropositive elements (–3.02 V vs. ENH), the thought of using Li insertion materials as the positive electrode of lithium accumulators was proposed. Therefore, due to the high reactivity of Li with H_2O, the transfer was made from an aqueous electrolyte to a non-aqueous electrolyte composed of lithium salts dissolved in an organic solvent. The Li-metal battery was introduced and marketed in 1985 by Moly-Energy. Despite their attractive performances (250 Wh/kg), the commercialization of Li-metal accumulators was short-lived due to their limited safety. In fact, some of these accumulators supplying mobile phones had to be withdrawn from the market after exploding in the ear of their users. The cause of this malfunction is associated with dendritic growth of Li caused by repetitive charging/discharging, which led to a short circuit inside the accumulator causing the explosion. This problem was known but not appreciated as it should have been due to the excitement generated by this technology with regard to the increased autonomy of mobile phones [TAR 01]. This was the end of the metallic lithium liquid technology.

To overcome the problems associated with dendritic growth, two strategies were rapidly proposed. The first strategy, proposed by M. Armand [ARM 79], involves replacing the liquid electrolyte by a conductive polymeric membrane which acts as a physical barrier between the positive and negative electrodes. All solid Li-polymer technology was initially developed by Hydro-Québec in the 1980s and was taken over by Bolloré 30 years later, which led to the first commercialization of this battery. Today, it supplies the "Blue Car" which crosses Paris as part of the urban project "Autolib".

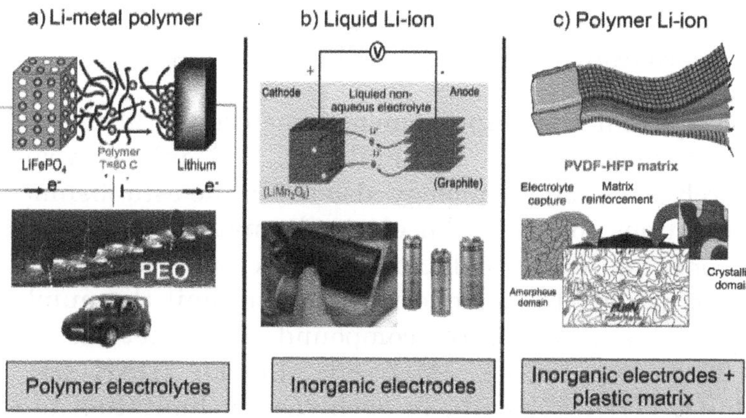

Figure 1.2. *Three currently used technologies based on Li-metal or Li^+ ions*

The second approach dates back to the 1980s and was initiated by Don Murphy [MUR 78] of "Bell Laboratories" (USA). He suggested replacing the negative Li electrode by an insertion compound that created a technology containing only Li^+ ions. This technology, devoid of Li-metal, was initially named "rocking-chair" accumulators, due to the back and forth motion of their ions between the positive and the negative electrode, and subsequently named Li-ion accumulators. Therefore, metal hydride-based systems with hygrogen ions functioning in an aqueous environment were duplicated and transferred to Li-ion systems in a non-aqueous medium. Li-ion technology was marketed in 1990 by Sony Energy [NAG 90]. Today, it is this technology which still prevails to power our mobile phones and has entered the electric vehicle market, and it is a serious candidate for grid applications (e.g. stationary storage for houses). A derivative of this technology was developed in 1995 [TAR 96] under the name Li-ion plastic technology, whose advantages resided in its flexibility and versatility with regard to its configuration, facilitating it to be adapted to the variable shapes of objects. The uniqueness of this technology is based on the use of a plastic composed of an electrolyte based on Li^+ attached to a solid polymer membrane based on poly(vinylidene fluoride-hexafluoropropylene) (PVDF-HFP).

Apart from the improvements, mentioned above, to the configuration of the accumulator, the initial years of research on Li-ion technology also led to the establishment of specifications for the ideal electrode material, namely good electronic and ionic conductors of a micrometric size with an open structure. These criteria of course limit the number of candidates and only three compounds have met the criteria for 20 years: $LiCoO_2$ [MIZ 80]; $LiNiO_2$ [DAH 91] and $LiMn_2O_4$ [THA 83] with, in each case, a limitation to their usable capacity of approximately 0.5 electrons per transition metal. This means that although $LiCoO_2$ theoretically has a capacity of 290 mAh/g, in reality only half, 150 mAh/g, can be utilized, due to the collapse of the structure caused by intersheet repulsion when too many Li^+ ions are removed. To overcome this problem of structural stability, chemists successfully resorted to cationic substitution. By partial substitution of cobalt by Mn and Ni within the metal sheet, lamellar oxides with capacities of 180 mAh/g called NMC [OHZ 01] were obtained and over time gradually replaced $LiCoO_2$ in commercial batteries. Many researchers had come to believe that Li-ion had reached its limits, bringing research to a halt. Some alternatives to the conventional approach, both with regard to the material and the reaction mechanisms, were needed to remove existing technological barriers, and favor technological breakthroughs.

These criteria for the ideal insertion material were questioned at the beginning of the 2000s with the arrival of nanomaterials for which the *Laboratoire de Réactivité et Chimie des Solides* (LRCS) [POI 01] was a pioneer. Along with the structure and composition, size was the third parameter; it was possible to recover many materials previously neglected due to poor ionic and electric conductivity or to uncover new reaction mechanisms that did not require lacunar compounds. The benefits offered by nanoscale and nanostructured materials to Li-ion technology are now so numerous that we will provide only the three best examples.

The first example is the transformation of $LiFePO_4$, by the nanoscale morphology/coating coupled approach [RAV 99], (an insulating material once abandoned) into one of the most sought electrode materials by battery manufacturers due to its abundance and low cost. It is ironic that this material (once insulating) is today the most efficient for power applications. The benefit of the nano/coating approach can be explained as follows: on the one hand, the nanoscale character of the particle reduces the travel time of an ion from the center to the surface of the material by a factor of 100 when the radius is divided by 10; on the other hand, the coating of a uniform layer of carbon on the electrode provides electron percolation for access to the whole active material. The two effects combined improve the kinetics of the electrode. Thus, today's Li-ion $LiFePO_4/C$ batteries, marketed since 2006 by A123, are still the preferred technology for electric vehicles and grid applications.

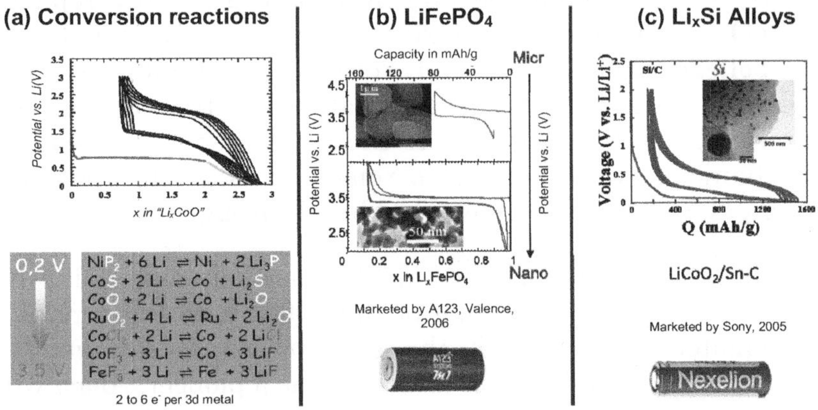

Figure 1.3. *The use of nanomaterials to transform $LiFePO_4$ insulators into attractive insertion materials a), to make the insertion of Li into Si possible b) and finally to develop new reaction mechanisms such as conversion reactions c)*

The second example shows how the transition from the nanometric scale helped solve the 20 year old problem regarding negative silicon or tin electrodes with lithium (Li_xSi_y) alloy reactions that caused significant changes in volume and consequently a loss of electronic percolation within the electrode [ANA 87]. Nanoparticles are in fact more resistant to fractures since the elastic energy that they store during deformation is not sufficient to initiate-propagate a fracture. In other words, for a nanoparticle, the volumetric strain relief required to initiate a fracture is not sufficient to counteract the surface energy associated with its propagation. Therefore, due to the strain relief provided by the nanoparticulate state, Si and Sn electrodes, combining nanometric character and carbon coating, were created [WIN 99]. They were marketed in 2005 (NEXELION technology with Sn) and in 2011 (Si technology by PANASONIC).

Finally, we recall that the transition to the nanometric scale also allows us to go beyond some well-established typical Li-ion insertion/deinsertion processes, by the discovery of a new mechanism reaction called "conversion reaction" [POI 01] which causes the breakage of bonds. According to this new mechanism, the electrochemical reduction of a binary oxide (CoO) by lithium forms a composite electrode formed from nanoparticles of metallic Co inserted into a Li_2O matrix which, due to its nanometric character, can be reduced during recharging. The reversible change from Co^{2+} to Co is associated with the absorption/desorption of two Li by 3D metal (instead of a maximum of one for insertion reactions) facilitating access to high capacity electrode materials. This surprising reactivity, previously unknown, proved universal, taking place regardless of the nature of the 3D metal or anion (F, O, S, N, P, etc.) with the possibility of adjusting: (1) the capacity of the electrode material which can reach up to 6 Li with 3D metals $(NiP_2 + 6Li^+ + 6e^- \rightarrow 2Li_3P + Ni)$ or even (2) the redox potential which can vary from 0.4 V to 3 V by increasing the ionicity of the MX bond. Thus, FeF_2 phases were studied as positive electrodes. Despite these improved capacities and good cyclability, it should be noted that unlike the previous technologies based on $LiFePO_4$ or Si, Li-ion systems with a conversion electrode are still under development due to the difficulties found with their energetic efficiency, which does not exceed 70%.

All the developments described above form the Li-ion technology of today, which, due to its assembly flexibility and performance, (~210 Wh/kg and ~800 Wh/L) which has doubled in 20 years, have conquered several markets. However, with regard to the evolution of this technology, we must ensure that it is also a success for electronic vehicles, hence the need to identify the critical points for this application.

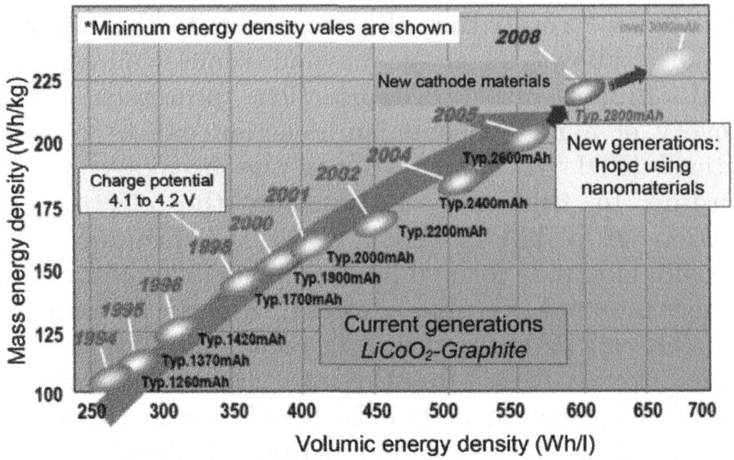

Figure 1.4. *Evolution of Li-ion technology since it was launched on the market*

1.2. Supercapacitors: recent developments

As for supercapacitors, as previously mentioned, they can deliver or recover large power peaks. However, they do have an energy density much lower than that of batteries [SIM 08]. The main challenge in this field is to increase the energy density of these systems, while maintaining the exceptional power performances. The energy E (J) given by the relation $E=1/2\ C.V^2$ where C is the capacity (F) and V is the voltage, there are several possible strategies to increase the energy density.

One line of research widely developed in the literature involves taking advantage of the different existing structural forms of carbon (Figure 1.5) to increase the capacity of supercapacitors by developing porous carbon structures whose pore size is best suited to electrolyte ions thus optimizing the capacity.

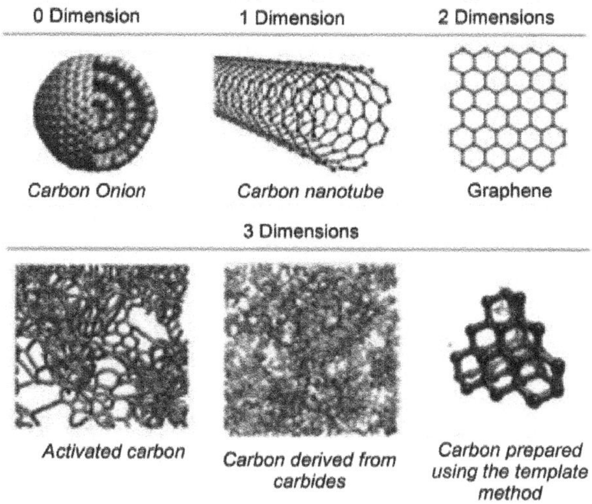

Figure 1.5. *The different carbon forms used in supercapacitors, from 1 to 3 dimensions [SIM 13] (source: Account for Chemical Research)*

Another strategy is to prepare pseudocapacitive materials by carbon replacement to improve capacity through redox surface reactions [SIM 08]. This can also be achieved by adding electroactive redox functional groups to the carbon surface [POG 12]. Here the challenge is to preserve the lifespan and find materials that function in an organic environment. Increasing the operating voltage also improves power and energy. Its aim, common with batteries, is to develop electrolytes with the highest possible electrochemical stability while maintaining sufficient ionic conductivity.

A final approach involves developing hybrid systems, which combine a battery electrode with a positive supercapacitor electrode. This idea is attractive as it increases both the capacity and voltage by carefully choosing materials for electrode. In practice, we come across several challenges including balancing of electrodes, limited lifespan of electrodes in batteries or even power limitations of these electrodes.

Figure 1.6 summarizes the evolution of the energy density of ultracapacitors over time according to the active materials used. We can see that the increase in the system's energy density requires the use of organic electrolytes. Hybrid systems are also of great interest but we must not forget that in the latter case, the increasing performance is usually done at the expense of power and cyclability, which are limited by the faradaic electrode.

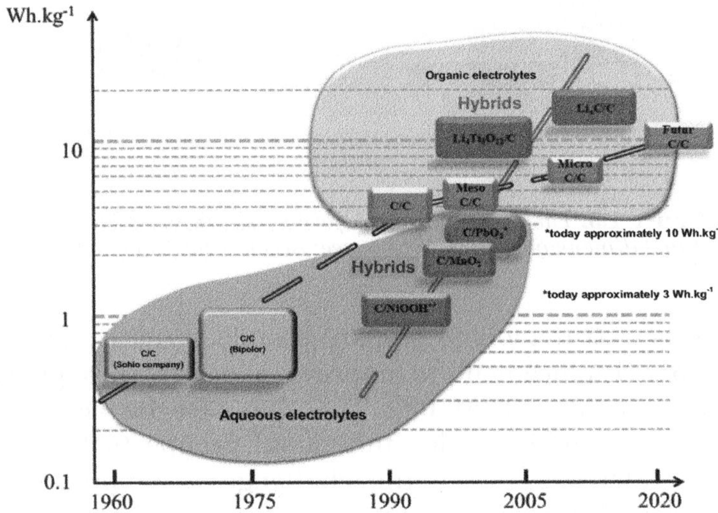

Figure 1.6. *Evolution of energy density of superapacitors over time, according to the different active materials used (source: l'Actualité Chimique)*

After having reviewed the state of the art and established the platform of knowledge required to appreciate the scientific challenges of this domain, we can now explain the logic, ambition and innovative nature of the RS2E's scientific program and reveal to the reader what we mean by research areas such as "Advanced Li-ion", "Eco-compatible storage", "New chemistries", "Smart materials" and "Capacitive storage".

2

Advanced Li-ion

2.1. Positive electrode materials for Li-ion technology

After 20 years of isolation, the research on Li-ion technology became highly diversified [ARM 08], especially with the arrival of nanomaterials, which opened up many opportunities that we will try to explain in this chapter. Technological breakthroughs still to be made on electrode materials, electrolytes and formulation to improve their performances, reliability and safety will also be discussed. This is a broad area including different accumulator components.

The quest for high-performing electrode materials is now fulfilled by $LiMO_2$-type lamellar compounds and polyanionic compounds [MAS 13] with the formula $AMXO_4$, with XO_4^{n-} being a polyanion, A an alkali and M a 3D metal. Their structure is an association of interconnected MO_6 octahedra and XO_4 tetrahedra within which are cavities occupied by Li^+ ions. This structure is not a very good conductor of electron delocalization because the MO_6 octahedra are isolated by XO_4^{n-} tetrahedra; these materials are, therefore, insulators and have been discarded as electrodes. With the transfer to nanometric scales, these materials became very attractive through the inductive effect of the polyanion with adjustable electronegativity, to adjust the redox potential $M^{n+} \rightarrow M^{(n-1)+}$ of the 3D element. Nowadays, studies focus on $LiFeBO_3$ [THA 05] and Li_2FeSiO_4 [LEG 01] phases with a detailed understanding of the mechanisms involved in the cycling of batteries. Special efforts have been made to improve the electrode kinetics by either introducing conductive polymers such as additives around the particles, or by developing new positive composite electrodes by nanoconfinement of active polyanionic species (vs. Li) in mesoporous carbon.

In the attempt to increase redox potential, the addition of a highly negative element such as fluorine has always been considered, such as fluorophosphates already mentioned by J. Barker and in particular phase $LiVPO_4F$ [NYT 05] due to its high voltage (~4.0 V). Through the production at low temperatures, a whole new family of fluorosulphates with the formula $AMSO_4F$ [HUA 09] was developed by RS2E, such as the $LiFeSO_4F$ phase with a triplite structure [REC 10], which inserts and disinserts lithium at a voltage of 3.9 V; this is still the highest redox potential for the couple Fe^{3+}/Fe^{2+} to have ever been reported for an inorganic compound. This sulfate chemistry has recently extended to cover hydroxy-sulfates [BAR 11], oxy-sulfates [RED 09] and even non-substituted $Li_2Fe(SO_4)_2$ sulfates, which have a redox potential of 3.83 V, and other pyrosulfates or even fluoropyrosulfates are now targeted. Whether sulfates or phosphates, these materials are also interesting since they are composed of abundant chemical elements and can be obtained from low-energy processes, thus obeying the concepts of sustainable development. However, their heavier weight, compared to that of oxides (due to their polyanion) stops them from competing with lamellar oxides in terms of performance, which makes them restricted to uses in large scale applications.

Thus, RS2E also became interested in $LiCoO_2$ and Li-NMC (acronym for Ni, Mn and Co) lamellar compounds, currently used in smartphone batteries. It has recently been shown that similar compounds containing excess Li (called lithium-rich NMC) may allow capacities greater than 250 mAh/g [ROU 14], i.e. an increase in autonomy of our portable devices by more than 30%. However, due to major challenges, in particular the drop in mean voltage of the accumulator as a function of the number of charge and discharge cycles, these materials have still not been commercialized, despite extremely active and competitive international research.

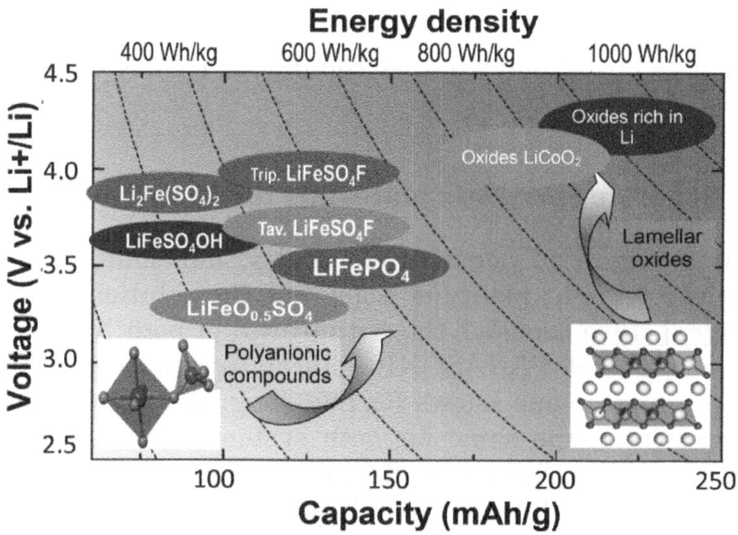

Figure 2.1. *Comparison of electrode materials studied by RS2E in terms of voltage, capacity and energy density (source: Chemistry of Materials)*

To shed light on these problems, RS2E researchers synthesized new families of lamellar oxides $Li_2Ru_{1-x}M_xO_3$ (M = Mn [SAT 13a], Sn [SAT 13b], Ti [SAT 14], etc.), which are structurally similar to NMC but with a simpler chemical composition. By combining XPS and EPR spectroscopies, it was possible to reveal the reversible participation of oxygen in redox processes. This, therefore, proved that the accumulation of cationic and anionic activity within the same material was responsible for this increased capacity. Moreover, out of the substituted compounds, only those substituted with tin showed a decrease in voltage. Comparative measurements using high-resolution microscopy on these materials substituted with titanium or tin revealed that this drop in voltage during cycling in these materials was due to the migration of cations and the capture of small cations (Ti) in the tetrahedra during cycling. This knowledge must now be applied to stabilize the voltage in Li-rich NMC phases, which is not at all obvious, and to better explore this new phenomenon to obtain materials with capacities far beyond what we could hope for five or six years ago.

2.2. Negative electrode materials for Li-ion technology

With regard to negative electrodes, nanomaterials have certainly brought plenty of hope, but the solution is clearly not ideal if we consider the withdrawal of both NEXELION technology and PANASONIC silicon technology from the market, and no one knows for how long. The reason for this withdrawal is the same for both systems and is related to their limited performances during cycling especially at high temperatures. This is one of the reasons why negative electrode compounds comprising an alloy material that reacts with lithium have been studied by RS2E. They stand apart from the studies in the literature, especially those with Si, that attempt to create elegant "Yolk-shell" or "pomegranate" architectures to optimize the Si electrode that are totally unrealistic [LUI 12] from an applicative standpoint. We prefer to focus our upcoming research on a chemistry aimed at (1) controlling the density and thickness of the SiO_x layer surrounding the silicon particle [LIU 14] and (2) grafting elastic or even conducting polymers [TOU 15]. The design of other promising compounds that can alloy with Li such as $Li_xTiSnSb$ [NGU 13] has also been studied. Finally, we note that through the confinement of the carbon matrix, high-performance SnO_2-based electrodes could be created, despite the problem of irreversibility, which is still preventing the use of such electrodes.

2.3. The question of electrolytes for Li-ion technology

The electrodes studied above function over a wide voltage range (0.1 V → 4.6 V) which requires the design of electrolytes that are perfectly resistant to oxidation and reduction. Liquid electrolytes and liquid ionic polymers have been considered, primarily to determine the best formulation by trial and error. The study of electrode/electrolyte interfaces and the impact of their formulation on the electrochemical performance of the negative or positive electrode material or even of the system is currently preferred at RS2E rather than the “design” of new salts or solvents.

3

Capacitive Storage

Studies on the topic of capacitive storage are oriented toward the increase in energy density of supercapacitors by working on carbon, pseudocapacitive materials and electrolytes.

3.1. Carbonated materials for capacitive storage

A *first research area* focuses on the development of porous carbon structures whose pore size is best suited to the electrolyte's ions to optimize capacity. One team at RS2E made a scientific breakthrough by showing that the electrolyte's ions can access pores smaller than the size of solvated ions and that in these subnanometric pores (<1 nm) the capacity of carbon was doubled [MAR 13]. Research has been carried out in particular to understand the mechanisms underlying this behavior. In collaboration with the transverse group "theory", studies combining experiments and modeling through molecular dynamics have been carried out. This collaboration between researchers and theorists, arising with the creation of RS2E, yielded top-class results, revealing a mechanism for the partial desolvation of ions [CHM 06, TSA 14] as well as ion exchange between the bulk electrolyte in the nanometric pores [MER 13] and the free electrolyte.

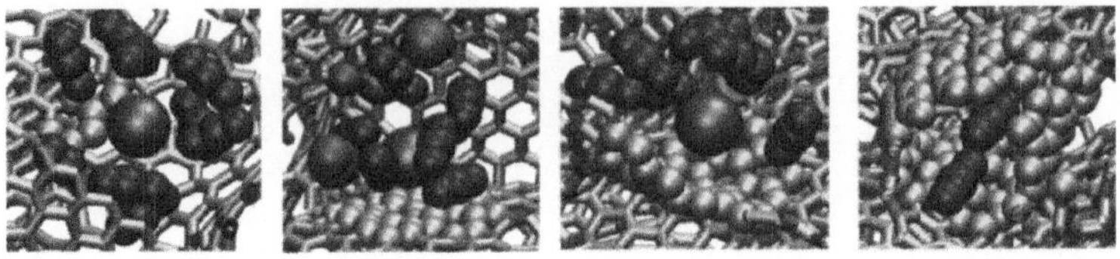

Figure 3.1. *Different types of environments encountered by ions (hexafluorophosfate in green, 1-butyl-3-methylimidazolium in red) and acetonitrile (blue) in nanoporous carbon electrodes (in turquoise) (source: Nature Communications). For a color version of the figure, see www.iste.co.uk / tarascon / storage.zip*

Based on these results, new methods for the synthesis of carbon with controlled hierarchical porosity in the nanometer range have been developed by the Institute Charles Gerhardt de Montpellier (ICGM) laboratory with remarkable electrochemical performances for an organic electrolyte [MER 12].

The functionalization of carbon by the addition of active redox groups (carboxylic acids, quinones, etc.) adds a redox component to the double layer capacity that increases the total capacity. In this area, the research carried out at the *Institut des Matériaux de Nantes* (IMN) laboratory provides a benchmark at the international level with a grafting technique based on the chemistry of diazonium [TSA 13]. If the grafted redox groups have the drawback of only being used in an aqueous medium, recent studies on this same group have shown the feasibility of active functionalization in an organic medium, which gives us a glimpse of real opportunities for the future [WEI 12].

3.2. Pseudocapacitive materials

Replacing carbon with pseudocapactive materials that store charges by fast surface redox reactions is another promising line of research for increasing the capacity of the electrodes, provided that it generates a cycle life much greater than that of batteries. RS2E is very active in this field, particularly on MnO_2, which is still the material of choice for these systems [LEB 14]. Research has also provided an insight into the capacitive storage mechanism in this material [GHI 14].

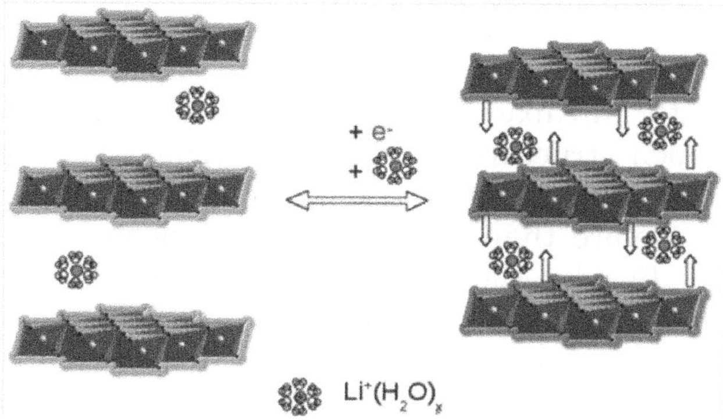

Figure 3.2. *Expansion-contraction of electrostatic origin of the MnO2 Birnessite sheet structure under the effect of the reversible intercalation of hydrated electrolyte cations [CAM 14] (source: Journal of Power Sources)*

The power performance of pseudocapacitive oxides, in particular manganese dioxides, is also limited by their low electronic conductivity. This is not the case for other oxides, particularly spinels, such as Mn_3O_4, Fe_3O_4 or even Co_3O_4 [GHO 12] and $HCoO_2$ [GOD 11] which have been explored. A disadvantage of these materials is that they all operate in an aqueous medium with a voltage limited to 1.5 V. The synthesis of nanostructured Nb_2O_5 using the sol-gel method has removed this barrier by providing an intercalation material for the Li^+ ions functioning in an organic medium based on pseudocapacitive reactions [PET 12]. If the capacity and power are high, the challenge is now to improve the cycle life.

The decoration of high surface area carbon with pseudocapacitive nanooxides is one strategy to increase the power density. Through this approach, we aim to maximize the electronic percolation and active surface area. Different nano- or mesostructured carbons can be formed. Precipitation or (electro) deposition of oxides in porous structures is preferentially treated with carbon whose meso- or macroscopic pore size and interconnections are large enough so that they are not blocked and allow the electrolyte to flow throughout the whole electrode. With these carbons (non-microporous), the increase in surface area, and consequently in capacity, are unfortunately limited. Carbon with an open structure, such as graphene, is preferred and graphene decorated with oxides such as FeOx provides a capacity of more than 100 F/g in an aqueous environment [AUG 13]. Besides the oxides, nitrides (and oxynitrides) of transition metals, such as TiN [GAO 14] or vanadium nitride (VN), [LUC 12] are being studied for pseudocapacitive applications. It turns out that the performance of these materials is very dependent on the chemical composition of the films and in particular the oxygen content. This composition is controlled through the use of specific synthesis techniques such as Physical Vapor Deposition (PVD), limiting the thickness of the films to µm, which is a hindrance except for applications in microelectronics.

3.3. Electrolytes for supercapacitors

Improving the energy density of supercapacitors can be done by increasing the operating voltage. This aim, the same as for batteries, is to develop an electrolyte with the largest possible electrochemical stability window while maintaining sufficient ionic conductivity. Aqueous electrolytes have a limited cell voltage of approximately 1 V due to reduction reactions and oxidation of water, and thus, electrolytes based on organic solvents have rapidly emerged. A typical electrolyte for the carbon-based supercapacitor Electrical Double-Layer Capacitor (EDLC) is a solution of $(C_2H_4)^{4+}$, BF^{4-} in acetonitrile, which has an electrochemical stability window of 2.7–3 V [GHI 11]. One alternative is to replace acetonitrile with propylene carbonate (PC), an alternative developed to find a substitute for acetonitrile without compromising power performance. The development of new electrolyte formulations that combine stability at high voltages and a suitable ionic conductivity is not easy, and there has been little innovative work in this area. The use of ionic liquids as salts dissolved in conventional solvents does not remove the solvent instability problem beyond 3 V, and the use of pure ionic liquids is tricky because of their low conductivity and extremely high melting point. Protic ionic liquids, with labile protons, are used pure and have an operating voltage of 2 V, but the capacity of these materials is still low due to the high viscosity of the electrolyte and the difficulty in their anhydrous synthesis [SIM 08]. Another possible alternative used by RS2E is based on the use of eutectic mixtures of ionic liquids, which helps lower the melting point over a wide temperature range [DEM 13, LIN 11]. By combining a eutectic mixture with carbon, whose surface is completely accessible, such as activated nanotubes or graphene, voltages above 3 V over a wide temperature range (–60°C to + 80°C) were obtained with capacities over 150 F/g in the case of activated graphene [LIN 11]. The cost of these electrolytes certainly remains high, but this approach is promising in the race to develop high-voltage supercapacitors.

3.4. Hybrid systems and middle-term goals

Finally, since there is always an attempt to increase performance, hybrid systems may offer some hope. These systems combine a battery electrode, such as a lithiated carbon graphite negative electrode, and a positive supercapacitor electrode such as porous carbon. The faradaic electrode provides a high capacity and the supercapacitor electrode maintains power performance. This approach is attractive because it can increase both the capacity and power by judiciously choosing the electrode materials. In practice, there are some challenges, such as the balancing of electrodes, the limited cycle life of the electrodes in the battery and the power limitations of these electrodes. Systems with lithiated carbon/carbon in an organic medium [TSA 13] or carbon/NiOOH [JME 09] in an aqueous medium are commercially available. RS2E research focuses on the development and study of low-cost systems in an aqueous medium such as carbon/MnO2 [SAF 15] or even hybrids Nb_2O_5/C or Ti_2C/C [BRO 10]. In summary, this is an interesting way to increase the energy density, but a cyclability and lifespan that is compatible with supercapacitor applications must also be maintained.

The aim of the research carried out within RS2E is, in the next 10 years, to double the energy density of supercapacitors beyond 10 Wh/kg, while maintaining a high-power density (> 10 kW/kg) and cyclability of several tens of thousands of cycles. Synthesizing carbon with a structure adapted to its electrolyte, the use of eutectic mixtures of solvents and the functionalization of carbon appear to be promising means of achieving these aims. This research will be carried out with the key support of multiscale modeling, not only to explain but also to predict the most suitable carbon structures. Hybrid systems combining faradaic pseudocapacitive electrodes or capacitive electrodes and battery electrodes are also likely to bring advancements due to the synthesis of nanostructured materials. It is only through an increase in their energy density that supercapacitors will become as essential as batteries in our everyday life.

4

New Chemistries

Despite the discovery of new materials and an improved anionic redox activity in integration processes, there is increasingly little space for improvement with regard to Li-ion technology. However, consumer demands are increasing, for example a desired autonomy of 500 km for the electric vehicle, instead of the current 180 km autonomy, for a lower cost. For grid applications, the selection criteria for systems include being maintenance-free and very low cost, regardless of the energy density. Ninety-eight percent of the grid energy storage is currently of hydroelectric origin, provided by the pumping of dams, yet the energy density produced by 1 kg of water falling by 1 m is 50,000 times less than that provided by a 1 kg battery [COM 14]. Besides safety, another challenge is the current need to develop rechargeable batteries in the context of sustainable development; this is not currently the case since recent studies, analyzing lifecycle, reveal that whatever the battery considered, 400 kWh of electricity and 80 kg of CO_2 emissions are required to manufacture a 1 kWh battery. This change in energy storage with new applications, different requirements and environmental constraints, makes no existing technology capable of meeting the demand alone, far from it. This situation, prone to diversification, caused prolific research into new batteries besides Li-ion, in order to meet new requirements. These new batteries, based on new chemistries, as we will see later, are nothing other than existing chemistries that were abandoned for various reasons.

Advances in reaction mechanisms through the gradual emergence of significant characterization tools such as electronic microscopy provide an insight into the reaction mechanisms of these older chemistries. This will hopefully remove the technical barriers that prevented their development in the first place. It is in this context that we are conducting RS2E research on Li-air technology as well as Li-S, solid-state Li, Na-ion, multivalent ions (Mg^{2+}) and redox flow with particular effort focusing on "eco-compatible storage" (a subject that will be covered later on). Activities conducted on these different battery chemistries within RS2E are described in the following sections.

4.1. Li-air technology

Is it possible to increase the energy density of lithium batteries to provide the desired autonomy for electrical vehicles or even grid applications? A glimmer of hope may actually come from Li-metal systems, and especially Li-air systems, which currently excite automobile manufacturers. Based on theoretical calculations, Li-air technology could provide energy densities of 3,500 Wh/kg, that is to say, approximately 15 times more than that provided by Li-ion batteries (Figure 4.1). However, to make such systems operational, many scientific and technological barriers need to be overcome, such as those related to energy efficiency and sustained cyclability, to name a few [LAR 15]. This is the purpose of the research jointly conducted by the RS2E and ALISTORE-ERI research networks.

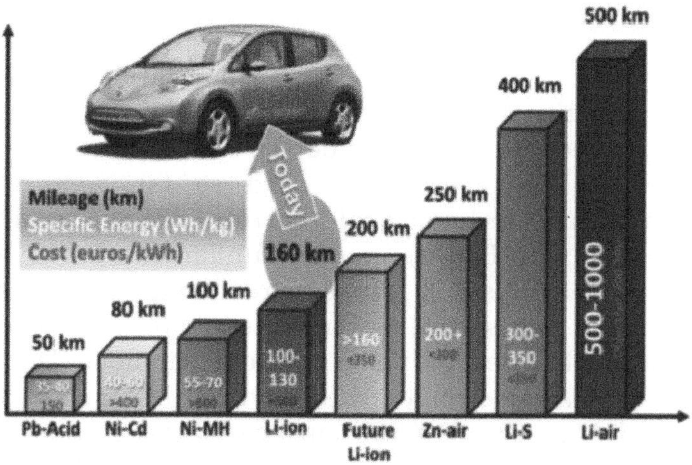

Figure 4.1. *Comparison of different battery technologies in terms of autonomy (source: Nature Materials)*

The Li-air battery uses Li metal as the negative electrode and an air electrode as the positive electrode, comprising a catalyst deposited on a highly porous carbon substrate. Unlike other systems in which the electrode materials are present within the battery, the Li-air couple is unique in that the active material of the positive electrode is not stored in the battery. On the contrary, the oxygen can be absorbed from the surrounding atmosphere and reduced on the surface of the electrode. Halfway between the fuel cell (which uses oxygen) and the lithium battery (which uses lithium), this new technology faces problems when both technologies are combined. These problems are related to the cyclability of the lithium electrode, control of the lithium/electrolyte interface, chemical stability of the electrolyte, corrosion of Li by oxygen and, last but not least, the dendritic growth of lithium during cycling at the anode (Figure 4.2).

With the nanostructuring of carbon-supported electrodes containing dispersed and nanometric α-MnO_2, Bruce *et al.* [BRU 12] showed the possibility of recharging such a system for several cycles with energy densities five to six times higher than for a Li-ion positive electrode, but with a polarization of 1 V. Reducing this polarization is the main challenge of this technology which involves the presence of highly nucleophilic superoxide O_2^- ions breaking down virtually all the electrolytes used so far. One way to overcome this technological barrier is to launch, in collaboration with Bruce's group, a fundamental study on the formation mechanism of LiO_2 superoxide and Li_2O_2 peroxide species in these systems. Contrary to established beliefs, we have shown that depending on the donor number (DN) of the solvent used as electrolyte, the growth of Li_2O_2 was through the solution or through the surface of the electrode. This emphasized the importance of the DN of the solvents used, which had been ignored up until then. The growth mechanism for Li_2O_2 is detailed in the following sentences. The oxygen O_2 arriving at the positive electrode is first reduced on the surface of the electrode through the capture of an electron to generate the superoxide O_2^- ion. Thereafter, the fate of this group will depend on the DN of the solvent. For solvents with a high DN, the highly dissolved Li^+ will be a weak acid producing unstable "Li-S-O_2^-" entities, which slowly disproportionate in the solution forming a uniform deposition of Li_2O_2 at the electrode. However, for solvents with a low DN, the undissolved Li^+ acts like a strong acid which reacts with the weak base O_2^- to produce a very unstable LiO_2, unlike the "Li-S-O_2^-" entity, rapidly decomposing into Li_2O_2. Therefore, the morphology of Li_2O_2 obtained depends on the growth speed, either fast or slow, with drastic consequences on the electrode performance. Solvents such as dimethyl sulfoxide (DMSO) with a high DN (DN $\approx$ 29) give much higher capacities than solvents with a low DN such as CH_3CN (DN $\approx$ 14). For the latter, the process of rapid disproportionation produces a Li_2O_2 film that blocks the whole surface of the electrode [OGA 06].

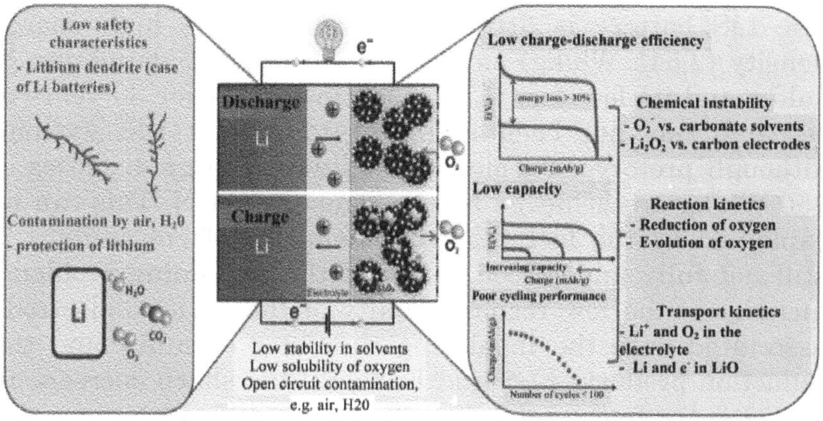

Figure 4.2. *Illustration of the main challenges associated with the development of Li-O_2 technology in an aprotic medium*

The effect created by this solvent, together with preliminary work (1) on a positive gold electrode [JOH 14] capable of reaching 100 cycles and later (2) on a TiC electrode [PEN 12] as well as (3) the benefit of using $LiNO_3$ as an additive [THO 13] in the electrolyte to create a protective SEI on the Li electrode, are encouraging advancements. However, the complexity of such a system is shown by the simple fact that after 6–7 years of research a good electrolyte could not be identified because Li-air technology is still far from being marketed. In addition to the electrolyte, studies aiming to replace the negative Li electrode and select the best catalysts, as well as mediator redox couples, are conducted by RS2E.

4.2. Li-S technology

Early work on the use of Li in the 1960s focused on the electrochemical Li-S cell. Its operation was based on the reduction of sulfur by lithium at the positive electrode to form a soluble polysulfide Li_2S_n chain ($3 \leq n \leq 6$) in the electrolyte, the final groups Li_2S_2 and Li_2S being insoluble. The LiS battery is attractive because of its high energy density (2,545 Wh/kg) as well as the natural abundance of sulfur and its low cost [LAR 15]. However, over 30 years of research has not been sufficient to market this system, although prototypes/demonstrators with energy densities of 350 Wh/kg are achieved by the company SiON. This can be explained by the fact that the chemistry of polysulfides is still not fully understood. Some of the remaining problems include (1) the low rechargeability of the positive electrode associated with the insulating character of sulfur and the reduction products Li_2S and Li_2S_2, (2) the sharp decrease in capacity due to the solubility of polysulfides which gives rise to a redox shuttle and, finally, (3) a Li-electrolyte interface that is even less well understood. The redox shuttle arises because the soluble polysulfides produced at the cathode are transported to the anode where they are reduced to polysulfides with lower *n*, which are again transported to the cathode where they are oxidized and so on.

Two strategies are currently being undertaken to minimize the redox shuttle. The first strategy is aimed at protecting the Li electrode by a $LiNO_3$ additive [UDD 13], using Li-S cells with liquid cathodes whose capacity is limited by the restricted solubility of polysulfides (~5 mol/L). The second strategy, which is the most common, is based on trapping the polysulfides at the positive electrode generated upon reduction [MIK 08]. One of the strategies used includes porous carbon (e.g. nanotubes) in which liquefied sulfur is impregnated. It is not necessary to say that the viability of this containment approach depends on the interaction between the polysulfides and the host structure. To overcome the low polarity of the carbon surface, researchers have also tried coating the carbon with a layer of polyethylene glycol, or replacing it with graphene, to take better advantage of the polarity of oxygen to trap sulfide ions. Similarly, C-S-(M-O) matrices have been made using oxides such as SiO_2 [JI 10], Al_2O_3 or even MOF [JI 11] as adsorbents each time with limited improvements, since all of these approaches decrease the electronic conductivity of the electrode. To solve this, L.F. Nazar [DEM 11, PAN 14] has recently proposed using a titanium metal oxide Ti_4O_7 in the mesoporous form. To date, these are the most spectacular results for composite electrodes loaded with 70% in sulfur. They can certainly be improved further by carefully choosing oxides to explore their electrical conductivity but also their acidic–basic character in studies conducted by RS2E via ALISTORE-ERI along with the development of the aqueous Li-S system [TAO 14].

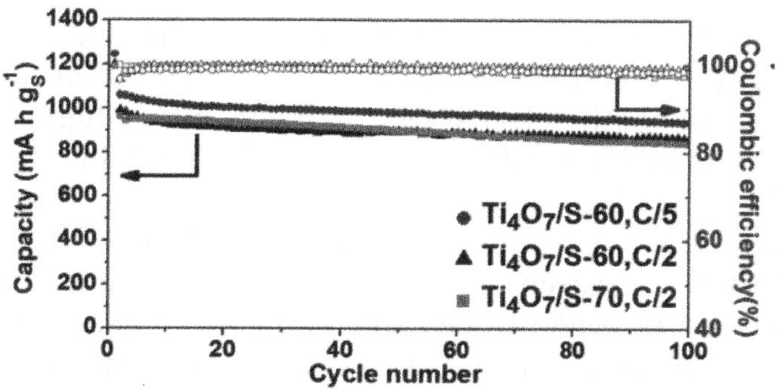

Figure 4.3. *Performance of Ti_4O_7-S composites in a Li-S battery (source: Nature Communications). For a color version of the figure, see www.iste.co.uk/tarascon/storage.zip*

4.3. Na-ion technology

One of the consequences of the success of Li-ion technology is that other potentially interesting chemical systems have been abandoned. Some of these systems, specifically those incorporating sodium as the exchangeable element, despite the obvious loss in terms of energy density compared to the Li-ion system, have many advantages from both an economic and ecological point of view and with regard to the intrinsic properties that control the efficiency of the electrochemical process. The inclusion of sodium in $ACoO_2$ lamellar compounds was studied by Claude Delmas [DEM 14] well before J.B. Goodenough studied the insertion of Li in $LiCoO_2$. With regard to ionic conduction, sodium β-alumina based compounds [DEL 80], or NASICON compounds, are more efficient than their lithium counterparts, which facilitates much faster transfer kinetics to be considered. Na-ion systems are therefore quite feasible, as demonstrated in 1999 through the assembly of the Na-ion system [YAO 67]: $Na_{3.7}Pb/NaCoO_2/PEO$. Thus, in 2010, ALISTORE-ERI began research on the Na-ion technology and was since joined by RS2E.

Just like its Li-ion counterpart, Na-ion battery will be much more efficient when the potential difference between the negative electrode and positive electrode, along with the capacity of both electrodes, will be very high. Active compounds at high voltages are already available; however, for the negative electrode, few compounds have been reported, if we exclude carbon and the Na_xPb_y alloy. Thus, our initial research focused on finding new negative electrodes. We identified the lamellar phase $Na_2Ti_3O_7$ capable of reversibly inserting a Na^+ ion at a voltage of 0.3 V, the lowest voltage reported [MA 93] for any sodium intercalation compound. With the cycling performance of this phase not possible to control, studies have been conducted in parallel (1) on carbon compounds with the discovery of a specific treatment that allows carbon with a capacity of 320 mAh/g to be obtained, which is constant over 100 cycles [SEN 11], and (2) on Li alloys with the possibility of creating a Sb electrode with a very good performance capacity during cycling and a rather impressive power performance [PON 13].

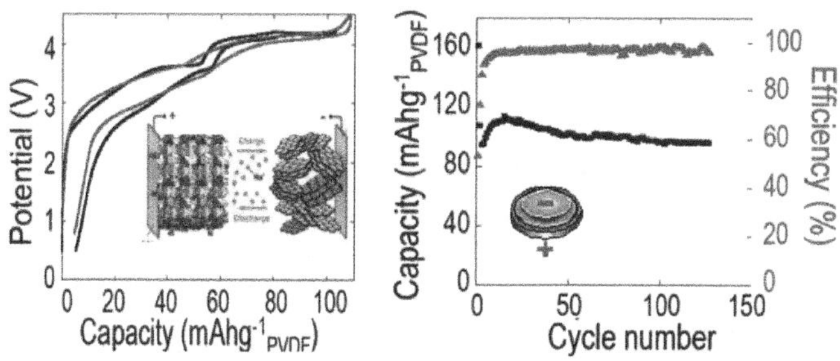

Figure 4.4. *Performance of Na-ion accumulators (C / NaPF6 / $Na_3V_2PO_4F_3$) carried out by RS2E and ALISTORE-ERI (source: Energy and Environmental Science). For a color version of the figure, see www.iste.co.uk / tarascon / storage.zip*

Studies conducted on electrolytes also proved to be reassuring [DAR 12] since they revealed that some details could also be used for sodium as well as Li, except that we had to use a sodium salt, generally less hygroscopic than its Li counterpart. The electrolyte 1MLiTFSI in ethylene carbonate-propylene carbonate-dimethyl carbonate (EC-PC-DMC) (45/45/10) with the addition of vinylene carbonate (VC) or fluoroethylene carbonate (FEC) is currently our reference. As for our research on positive electrodes, particularly on the possibility of polyanionic and lamellar compounds, we managed to identify phase $Na_3V_2PO_4F_3$ [PON 12] as a possible cathode for the insertion of Na due to its high voltage and stability in air, as opposed to lamellar phases which have a greater capacity at lower voltages but react with air.

Provided with all this work, RS2E researchers were able to assemble coin cell batteries based on the electrochemical chain Na_xC/1MLiTFSI/$Na_3V_2PO_4F_3$ [PON 13] in EC-PC-DMC. The performances obtained allowed us to confirm that Na-ion technology was feasible; it still remained to be known whether there was an advantage in terms of cost, safety and performance compared to its Li-ion counterpart, hence the need to move to the next level, which would not have been possible without the RS2E. Thus, in July 2013, the decision was made to put together a joint task force between

RS2E's academic researchers and CEA (French atomic energy and alternative energies commission) to assemble Na-ion 18650 cell prototypes (1) to demonstrate the concept but also (2) to assess Na-ion technology in terms of cost and performance and, of course, (3) to develop fully operational French Na-ion batteries. The first prototypes were assembled in June 2014 and we suggest readers refer to the upcoming book comprised in this set (by J.-M. Tarascon, P. Barboux and R. Palacin), published by ISTE and dedicated to the Na-ion, in which the results will be presented.

It is not necessary to say that this work goes hand in hand with fundamental studies on transport mechanisms, studies on interfaces, as well as prospective studies on other systems with multivalent cations such as Ca^{2+} and Mg^{2+}.

4.4. Redox-flow technology

The creation of RS2E coincided with a renewed global interest for redox-flow systems, although less publicized than Li-air systems, but real. Considerable effort had been made in China, Japan, United States and Europe, in particular with the English company Plurion, whereas French efforts in this domain were at best sporadic, if not non-existent. Therefore, it was strategically important for France to position itself in this domain, and through RS2E we could quickly set up a critical mass group whose mission was not only to position itself in this sector but also to become innovative.

The development of redox cells with circulating flow (or redox-flow) is not new since it dates back to 1968 with the invention of the zinc-chloride (Zn-Cl) battery. Half-way between fuel cells and batteries, these systems involve two soluble circulating redox couples separated by an ion exchange membrane. These redox couples, stored in two different reservoirs, called the catholyte and anolyte, are continuously injected using a pump inside the cell where they are reduced or oxidized to store or release energy [YAB 12]. The advantages of such systems are mainly due to the following possibilities: (1) power–energy uncoupling (number and size of cell versus concentration and volume of catholytes and anolytes), (2) low-cost extension and (3) to respond to high demands for autonomy dictated by renewable energy applications [KOM 11]. Redox-flow batteries are more relevant when the energy required is greater than tens or hundreds of kWh. Many redox accumulators exist, the most familiar being the all-vanadium system (1.7 V) which is based on the four possible oxidation states of vanadium (V^{2+}/V^{3+} – V^{4+}/V^{5+}) [BAR 89]. They differ by their chemistry (nature of redox couples) and configurations (one vs. two reservoirs) but share the same type of liquid electrolyte, which limits their output voltage per cell and contributes to the low energy densities of these systems also penalized by the associated machinery (pumps and reservoirs) and the robustness and selectivity of membranes that limit the lifetime.

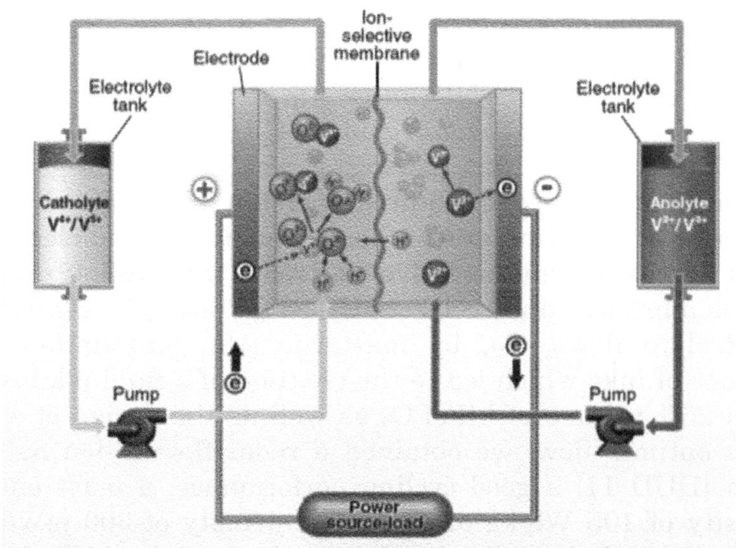

Figure 4.5. *Diagram of redox-flow battery based on the use of V^{4+}/V^{5+} and V^{2+}/V^{3+} couples such as anolyte and catholyte, respectively (source: Science)*

The studies undertaken by RS2E differ completely from previous work as they initially tried to develop a Li-ion redox-flow system, followed by work recently undertaken by a group at MIT [DUN 11]. This new concept, Li-ion redox-flow, which uses inks composed of powders of electrochemically active materials in suspension in the electrolyte, and not solutions, has many advantages in terms of (1) performances (higher energy densities associated with a higher concentration of active species (10 times higher (×10) than in an aqueous system) and an output voltage three times higher) and (2) lifetime (conditions of use much less restricted with regard to the separating membrane due to the absence of a corrosive medium and a specific ion-counter ion).

In addition to our good knowledge of the electrochemistry of this system due to its similarity to the Li-ion system, including the use of the same positive electrode materials (e.g. $LiMn_2O_4$, $LiCoO_2$), negative electrodes (C6, TiO_2 and $Li_4Ti_5O_{12}$) and electrolytes, we still had to learn about macro- and microfluidics and RS2E allowed us to obtain this through relations Solvay-CNRS's Lab of the Future, CIRIMAT (*Centre Interuniversitaire de Recherche et d'Ingenierie des Matériaux*) and LGC (*Laboratoire de Génie Chimique*). Due to this pairing, it was possible to gain the platform of knowledge required to create innovative methods in terms of (1) formulation of ionic, conductive and electronic inks for positive and negative electrodes [SKY 86], (2) design of membranes and cells [DUD 11] and (3) control of electrolyte flows. So, by mastering the formulation and science of inks which led to the creation of a fluid ink loaded with 25% volume of $LiFePO_4$ as well as the design of a cell with optimal flow, we obtained a redox-flow Li-ion battery with [DUD 11] a good cycling performance, a mass energy density of 105 Wh/kg and a power density of 300 mW/cm^2, surpassing the results previously obtained by MIT. Using the same cell, another noteworthy result was the possibility of cycling a Si suspension, with excellent capacity retention and a good coulometric yield, although the cycling Si of electrodes had been problematic for a long time for Li-ion technology. Thus, it may be possible to rethink the design of such electrodes.

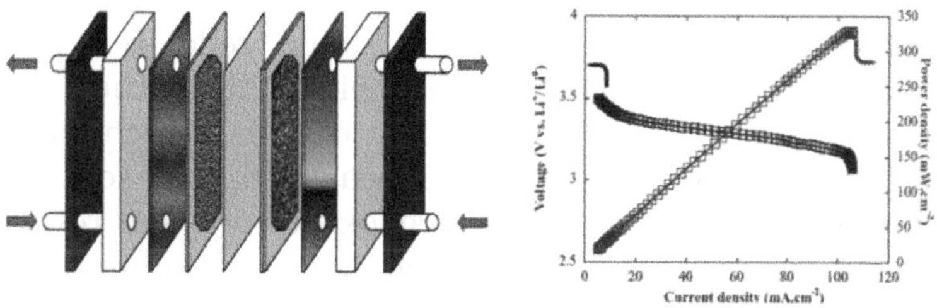

Figure 4.6. *a) Diagram of redox-flow cell using $LiFePO_4$ and $Li_4Ti_5O_{12}$ suspensions as the positive and negative electrode, respectively, and b) curves of discharge as well as power density of the system (source: Journal of the Electrochemial Society)*

Good performances can be obtained with the non-aqueous Li-ion redox-flow system; however, a technical discussion of these results with RS2E's Industrial Partners immediately concluded that the economic viability of the system (for stationary applications) needed to focus on aqueous systems (for reasons of safety, toxicity and cost). This is the direction our research is currently taking.

Besides the transition to an aqueous system, another approach pursued within RS2E to lower the cost of such a system is to remove the exchange membrane, which is costly. The underlying idea behind this work comes from an observation of nature, and the fact that when two rivers with transparent and muddy waters meet, it takes dozens or even hundreds of kilometers for the waters to mix. The distances covered by the flow of the catholyte and anolyte in parallel within a redox-flow cell are much lower; there was great reason to believe that the catholyte and anolyte would not mix at all. The assembly of a microfluidic redox-flow cell, using $LiFePO_4$ and $Li_4Ti_5O_{12}$ inks circulating in parallel, has effectively shown, through optical measurements and X-ray diffractions, the absence of mixing between two flows. However, the electrochemical response of the cell was distorted due to the contact of positive and negative species at the interface. To solve this problem, we simply introduced a third fluid (electrolyte only) between the anolyte and catholyte flow, thus achieving a Li-ion redox-flow system with a liquid membrane, thereby eliminating the physical membrane. Optimizing this concept of cells, which gave rise to two patents, is now heavily researched, and so does our desire to use currently passive redox flow system reservoirs as active components of the system through the use of redox shuttles.

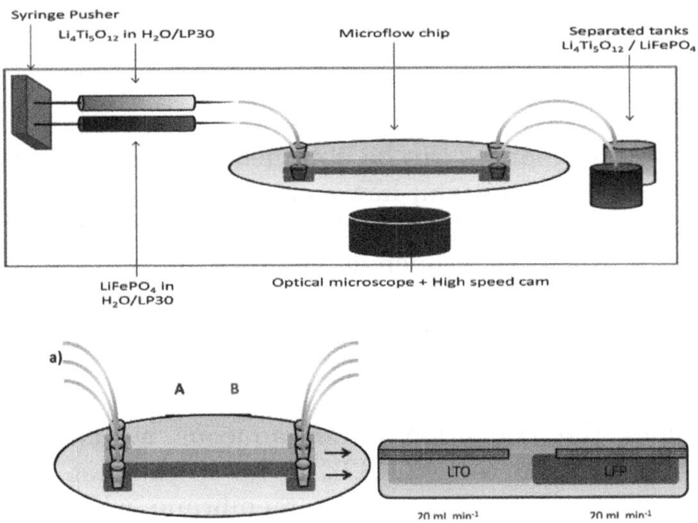

Figure 4.7. *Diagram of a redox-flow cell without a porous membrane*

4.5. All-solid state batteries

Twenty years ago, France held a leading position in this field, with strong teams; however, these activities have mostly disappeared, and are now scattered, except in Bordeaux. The "all-solid battery" is currently attracting renewed interest due to its safety. It requires highly conductive ceramic membranes which are also of interest with regard to Li-air or Li-S technology for which liquid electrolytes are a major problem. This is a strategically important area of research for RS2E, a choice reinforced by the dominant positioning of some industries in this field, such as Toyota. Consequently, an "all-solid battery" group was created within the RS2E, whose objectives are centered around: (1) the development of innovative assembly methods using Spark Plasma Sintering (SPS) or flash sintering [YOU 13], (2) the development of original approaches for the fabrication of low-impedance electrode/electrolyte interfaces with a high degree of elasticity to prevent delamination during cycling, currently the Achilles' heel of these batteries, and (3) the research of new ionic conductors to fabricate better conductive ceramics of Li^+ or Na^+ ions.

There are two divisions that currently exist with regard to conductive ceramics: first, there are advocates of oxides due to their easy control and non-reactivity to air – even though their ionic conductivity cannot exceed 3×10^{-3} S/cm for $Li_{1.3}Ti_{1.3}Al_{0.3}(PO_4)_3$ (LATP). Second, there are advocates of sulfides. These materials have already been heavily studied in the past, but have since then been neglected due to their difficult control and high reactivity to moisture. Recently, they have regained interest following the work of Kanno [HAM 12], who showed that the phase $Li_{10}GeP_2S_{12}$ actually had a conductivity of 12 $mS.cm^{-1}$ at 27°C, an extremely high value equal to or higher than the conductivity of practical liquid organic electrolytes used in Li-ion batteries. Note that the only ion conductive ceramic of Li^+ ions currently marketed by the OHARA Corp. is the phase $Li_y Al_xTi_{2-x}Si_yP_{(3-y)}O_{12}$. In this context, parallel research is carried out in this network on (1) sulfides (Li_4SnS_4 or even other phases derived from the compound of Kanno), (2) NASICON oxides but also S- and Cl-based compounds such as the phase Li_6PS_5Cl, which belongs to the family of Argyrodite Li_6PS_5X ionic conductors (X = Cl, Br, I) [ABO 11]. The novelty here was the opportunity to prepare this phase at room temperature and achieve the best conductivity ever reported at room temperature; α = 1.33 × 10^{-3} S/cm. Using this ionic conductor, an all-solid $LiCoO_2/Li_6PS_5Cl/Li_4Ti_5O_{12}$ battery with a reversible capacity of 27 mAh. g^{-1} of active material over at least 10 cycles and a coulombic efficiency close to 99% was obtained.

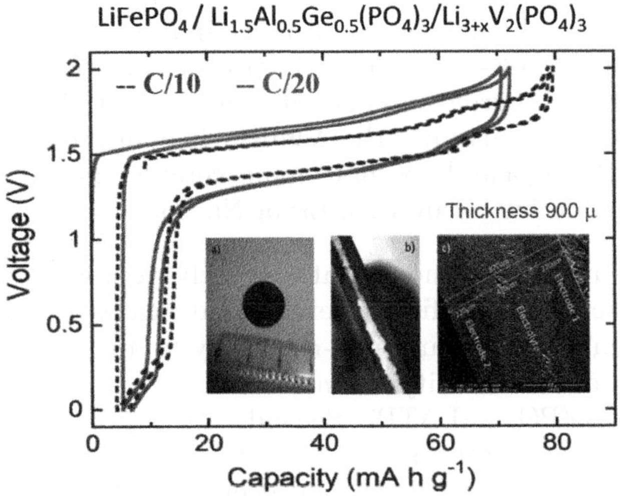

Figure 4.8. *Performance of a totally solid $LiFePO_4$ / LAGP / $Li_{3+x}V_2(PO_4)_3$ battery with an insert of the cross-section of the battery to illustrate the thickness of each component as well as the quality of interfaces (source: Advanced Functional Materials). For a color version of the figure, see www.iste.co.uk / tarascon / storage.zip*

The possibility of assembling an all-solid battery by SPS was demonstrated within the RS2E network in 2011 with the manufacture of an all-solid $LiFePO_4/Li_{1.5}Al_{0.5}Ge_{0.5}$ $(PO_4)_3/Li_{3+x}V_2(PO_4)_3$ battery that can be cycled at 120°C for more than 100 cycles with currents of 80 mA/cm^2 [YOU 13]. By varying the assembly methods, an all-solid symmetrical monolithic Na-ion battery $Na_3V_2(PO_4)_3/Na_3Zr_2Si_2PO_{12}$/

$Na_3V_2(PO_4)_3$ of 560 nm thickness was assembled in a single stage in 10 min by flash sintering at 900°C [KAM 11]. Upon cycling, this battery exhibits interesting performances for currents less than C/10 but rapidly degrades with currents of C/10 and beyond; the reason being mainly due to resistance at the interfaces increasing during cycling.

This new insight into the all-solid battery has facilitated a wider range of assembly techniques to be developed. This alternative technology is full of promise, but its potential advantages are still, for now, largely eclipsed by many technological barriers limiting its performances in terms of cycling and overall performance. Finding the best ionic conductors, whether Na or Li, is essential since the problems associated with Li-air or Li-S technologies would be largely simplified by the existence of high ionic conduction membranes that can physically separate the positive and negative compartments, and thereby solve the problems related to redox shuttles associated with polysulfides for example (case of the Li-S technology). Better membranes would also facilitate the development of aqueous systems similar to those already started by S. Visco [BOU 13]. However, the best ionic conductors today do not seem to guarantee success or ensure the feasibility of all-solid batteries. All-solid batteries based on the $Li_{10}GeP_2S_{12}$ phase have a better ionic conductivity than our liquid electrolytes that still have the same issues with interfaces. The message is clear; the interface should be at the center of ACTIONS in our future activities. It is necessary to overcome the resistance in transport at the electrode/electrolyte interface. This is the direction Toyota took by proposing to inject a $Li_{1.1}(Nb_{0.5}Ta_{0.5})_{0.9}O_{3-\delta}$ phase with a high dielectric constant and permittivity to considerably decrease the resistance to transport of charge at the interface. With regard to the elasticity at the interface, one solution may involve elaborating heterostructures with a concentration gradient at interfaces which takes the change in volume at the interface in microns rather than nanometers. Making such heterostructures is common in microelectronics (lasers, etc.) and is feasible for particles/powders as recent research by Y-K Sun on lamellar particles has revealed. Materials belonging to the NASICON family are particularly suitable for this as they provide a highly versatile range of cationic and anionic substitutions within the same structure.

5

Eco-Compatible Storage

Finding new compounds is one thing, but turning them into useful materials in the context of sustainable development requires new production strategies. To do this, we have deviated from energy-intensive ceramic methods, using high temperatures to ensure the diffusion of the reactants necessary for the growth of new phases, toward low-temperature methods known under the term *"chimie douce"*, or soft chemistry, and using synthesis in solution. We will revisit well-known methods such as the hydrothermal or solvothermal methods or even explore new routes with regard to the ionothermal approach as well as bioinspired/biomimetic approaches or new concepts such as renewable electrodes, which all fall within the framework of sustainable development. However, designing a rechargeable battery also requires the following to be taken into account (1) consideration of the lifecycle cost of materials and (2) recycling problems, since the number of batteries will surpass 10 billion in 5–10 years from now. These different aspects, some of which are new to RS2E, are discussed in this chapter.

5.1. Ionothermal synthesis

RS2E was a pioneer in the introduction of ionic liquids in inorganic synthesis [LAL 14] as a reaction medium and structuring agent. In addition to lowering the usual preparation temperature of electrode materials from 400/500°C, another benefit of this synthesis is the ability to adjust the structuring power of the ionic liquid, which depends on the nature/composition of the cations and anions composing it, and to control/direct the growth of new phases or even to prepare monocrystals for detailed studies of transport/diffusion properties. A new family of fluorosulfates with more than 20 members with varying physical and electrochemical properties has been discovered. One of them, LiFeSO4F, with a triplite structure, has a voltage of 3.9 V versus Li^+/Li. In addition, the ionothermal synthesis together with solvants (water or alcohol) and in the presence, or not, of surfactants allows micellar systems to be prepared that can adjust the texture and morphology of powders.

Due to the chemical richness of ionic liquids, whose number is virtually infinite, ionothermal synthesis provides a myriad of opportunities to stabilize new inorganic phases, impacting many fields of applications beyond that of electrochemical storage. This synthesis is particularly useful in the case of materials with high reactivity/solubility in aqueous media. We are, therefore, trying to extend it to the development of other types of polyanionic materials based on Fe or Mn (fluorocarbonates, orthosilicates and borates), or even pyrosulfates, fluoropyrosulfates of 3D metals, previously unknown.

5.2. Bioinspired synthesis/approach

If ionic liquids meet certain precepts of "green chemistry" and sustainable development, they do not allow the development of nanoscale or solid materials at room temperature as we might wish. To fill this gap we have tried, in collaboration with biologists, to synthesize inorganic materials in a biological or enzymatic medium, therefore, involving the direct use of bacteria or genetically modified viruses. Initially, as in previous studies by Belcher *et al.* [VIS 13], we tried using genetically modified viruses as "active" template agents to prepare textured inorganic materials. The complexity of this approach, which has produced inconclusive results, has shifted to "biomineralization", with many examples in nature including the development of highly textured silica shells produced by well-known unicellular algae, the diatoms. With this in mind, we moved toward the use of other microorganisms simpler than viruses, such as bacteria.

At first, we were able to show, despite difficulties in reproducibility, that it was possible to prepare nanometric $LiFePO_4$ at 60°C through the addition of ureolytic bacteria – *Bacillus pasteurii* – and appropriate precursors. Thereafter, we chose a strain of anaerobic ferroxidant bacteria (*Acidovorax* sp., $BoFeN_1$), capable of oxidizing Fe^{2+}. The distinct trait of this oxidation is that its product (lepidocrocite α-FeOOH) preferentially crystallizes between the two membranes of the bacterial cell wall (the periplasmic space), that is to say, in a very confined space. The precipitated nanoparticles are then arranged to form a shell matching the shape of the bacterium. A short heating in air at 700°C eliminates the organic/living part of the complex, whereas FeOOH is transformed into α-Fe_2O_3 hematite. These changes do not affect the initial organization of the particles, and so we obtain hollow bacteriomorphous shells composed of hematite nanoparticles [TAR 10].

Thanks to this unique alveolar porous texture induced by bacteria, these textured hematite samples display interesting electrochemical properties in terms of potential behavior, when used as electrode materials. Such electrodes can be charged and discharged 50 times faster than those using non-textured materials.

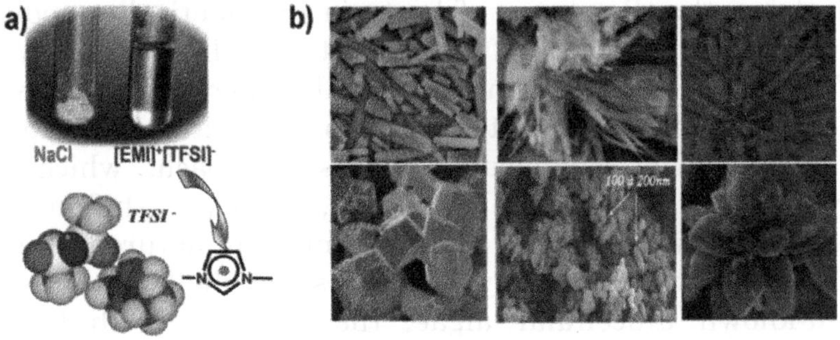

Figure 5.1. *State of two salts at ambient temperature. The solid state for NaCl is compared to the liquid state for EMI-TFSI in a). The diversity of morphologies that can be obtained for a compound by using ionic liquids with different cations and anions b). For a color version of the figure, see www.iste.co.uk/tarascon/storage.zip*

Figure 5.2. *SEM images of bacteria/FeOOH composites (left) and α-Fe_2O_3 bacteriomorphes (center) +Image TEM (right) showing a cross-sectional view of α-Fe_2O_3 bacteriomorphes (source: Energy and Environmental Science)*

These studies demonstrate the convenience of biologically assisted synthesis for the development of electrode materials. However, this eco-friendly synthesis approach is still dependent on our ability to control the physiological conditions of bacteria and to accelerate their reaction kinetics. These are the kinds of challenges to which we will devote our future activities in collaboration with our biochemist colleagues, while trying to maximize the anaerobic metabolism of iron-oxidizing bacteria to charge lithium-ion (Li-ion) batteries.

5.3. Organic electrodes for "green" Li-ion batteries and more durable batteries

At this point, note that current Li-ion technology is based on inorganic materials whose abundance and cost are often challenged. Would it not be better, henceforth, to embark on research that will prevent future shortages of electrode materials such as those from fossil fuels [NAM 06]? In this context, the development of rechargeable Li-ion batteries that are easy to recycle and capable of being integrated into a lifecycle compatible with the concept of sustainable development, is another RS2E activity. All energetic processes in living organisms are based on redox couples involving organic molecules: why not try to use them in our batteries? This idea is not only proposed but also validated by researchers at RS2E through the development of renewable organic molecules with carbonyl or carboxylate groups which are electroactive versus Li [MOI 14].

Among these is the quinone redox system, $Li_2C_6O_6$, [TAR 08] which can be synthesized by natural precursors with an electrochemical activity versus Li^+. This molecule can react reversibly with $4Li^+$ per unit formula at an average voltage of ~ 2.5 V; which leads to a gravimetric energy density almost twice as high as that for current commercial $LiCoO_2$ electrodes. With regard to the negative electrode, studies were also crowned a success by the discovery of carboxylate-based organic molecules (such as dilithium terephtalate ($Li_2C_8H_4O_4$)) capable of reversibly reacting with two lithiums per unit formula at a voltage of 0.8 V [POI 11]. These two examples confirm the possibility of using biomass to produce electroactive organic electrodes, despite the fact that many studies still need to be carried out for this concept to become a product. For example, the identification of new lithiated organic molecules [CHE 08] capable of releasing Li at high voltages, which is vital to create eco-friendly Li-ion rechargeable batteries. Also essential is (1) the development of hybrid structures produced by grafting of electroactive organic molecules onto nanometric oxides to prevent solubility during prolonged cycling or (2) improve the electrode kinetics. It is, of course, clear that this type of research, given the long-term benefits that this new concept could provide with regard to the CO_2 footprint, is currently being carried out within RS2E.

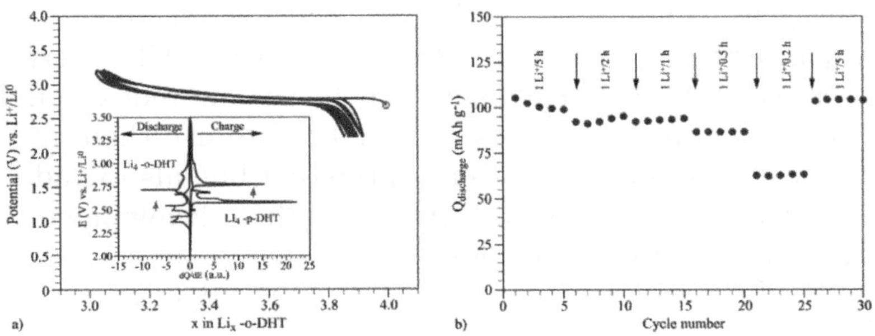

Figure 5.3. *a) Galvanostatic curve (voltage-composition and derivatives) for Li_4-o-DHT showing the increase in voltage compared to Li_4-p-DHT and b) retention of associated capacity according to the regime imposed (source: Applied Materials & Interface)*

Among the most notable results, there is evidence of the key role played by the terephtaloyl unit [ARM 09] by substitutions guided by theoretical calculations to obtain lithiated compounds such as dilithium (2.5-dilithium-oxy)-terephthalate (Li_4-*p*-DHT) capable of generating an operating voltage of at least 3 V versus Li^+/Li, which begins to make the development of an all organic Li-ion battery prototype possible.

5.4. Recycling and LCA

Given the increasing number of rechargable Li-ion batteries produced annually, it seemed important to focus on their end use, their recycling and also their eco-design, hence the reason for launching a group on recycling – Life-Cycle Assessment (LCA). The lack of expertise among the founders on these aspects of recycling led us to request the collaboration of *Laboratoire d'Électrochimie et de Physicochimie des Matériaux et des Interfaces* (LEPMI) (P.X. Thivel) and *Institut de Recherche de Chimie Paris* (IRCP) (A. Chagnes) whose mission is to develop and improve the recovery and recycling processess of energetically and strategically interesting materials and, on the other hand, to apply eco-design methods to future batteries. Due to the compatibility of these teams, especially with, on the one hand, competencies in the engineering of processes applied to recycling which is developed by LEPMI, and on the other hand the physicochemistry of separation processes developed by *Chimie ParisTech*, initially applied to the hydrometallurgy of primary resources, RS2E now has a team to match the ambition of its projects – projects that are currently centered around electro-assisted selective leaching for recycling of future generations of Li-ion batteries by coupling a chemistry approach with process engineering.

6

Smart Materials

Besides batteries, there are many applications that are based on “redox” reactions, such as photoelectrolysis. Therefore, some materials, such as TiO_2, play a key role in photovoltaics, photoelectrolysis and the electrochemical storage of energy. The aim of this chapter will, therefore, be to discuss the exploitation or development of smart materials with, due to their redox behavior, modifiable optical or thermoelectric properties that allow ingenious bifunctional architectures to be built from the same material. This is a new topic created during the launch of RS2E, requiring a systemic approach that has led us to expand the list of our partner labs including membership of IRCICA/IEMN (*Institut de Recherche sur les Composants logiciels et matériels pour l’Information et la Communication Avancée*) whose globally recognized research on microelectronics and nanostructuring was crucial for the development of devices with innovative architectures. Research on the photonics/storage coupling and the capacitive/faradaic coupling today are its strong and original points.

6.1. Photonics of insertion materials to create photo-rechargeable batteries

Early studies published by Gerischer and Tribusch in the 1970s and early 1980s [REN 13], concerning the electrochemistry of semiconductor electrodes under illumination, revealed the possibility of (1) extracting protons from the surface using a TiO_2-B electride under illumination and (2) inserting or deinserting copper ions into transition metal sulfides. Then, 10 years later, a photo-rechargeable battery using the lamellar $CuFeTe_2$ was reported [BON 13]. Why not take advantage of this phenomenon with Li^+ insertion compounds and develop photo-rechargeable batteries? This has been successfully tried by RS2E teams. As the mechanisms of photo-insertion and photo-deinsertion are confined to the surface of the material, the first step was to control not only the morphology but also the growth of TiO_2 particles to obtain a high state of division [BET 84]. This, therefore, focused on TiO_2 anatase because already knowing its optical and photoelectrochemical properties, it intercalates lithium at a high redox potential of 1.85 V versus Li^+/Li. Note that to properly understand the results when in the presence of a semiconductor (TiO_2 photo-anode), electrons (major carriers) migrate toward the center of the particle, unlike electron holes, which migrate toward the surface (minor carriers). Holes created due to the absorption of photons will react due to the Ti^{3+} present in phases Li_xTiO_2 and cause the deinsertion via a charge compensation mechanism. On this basis, a photo-rechargeable lithium/TiO_2 battery of approximately 1.8 V (patent pending) has been developed; the same concept was also developed by a Korean group who reported to have achieved photo-rechargeable batteries [NOM 95], the battery this time being a redox flow battery and not a Li-metal battery. These results are very innovative and promising, but of course need to be highly optimized. These improvements, currently underway, are designed to bypass (1) intrinsic absorption limitations of TiO_2 due to its initial bandgap of 3.2 eV and (2) the low electronic conduction under illumination does not allow a highly efficient collection of electrons. It is in this context that the chemistry of sulfides, intercalation materials with low bandgaps and good electrical conductivity, may become interesting. Many groups [PAT 13] are interested in the increased absorption of di-chalcogenide monolayers (MoS_2, WSe_2, etc.).

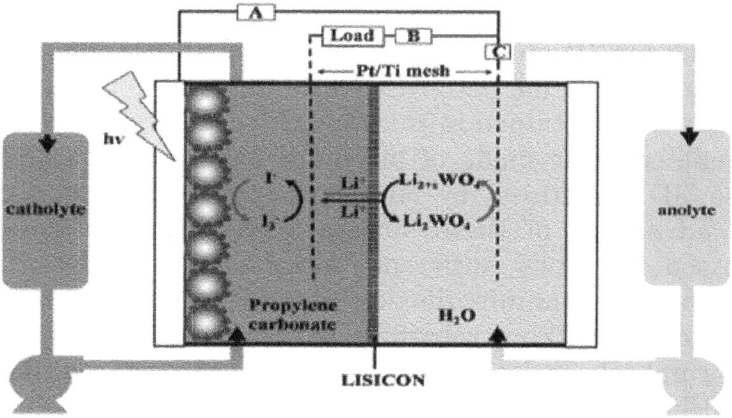

Figure 6.1. *The solar rechargeable redox flow battery based on Li_2WO_4/LiI couples from [YAN 13]*

6.2. Micro-energy sources

Currently, it is the lack of miniaturized efficient energy sources that restrict the development of emerging technologies in the field of biomedicine and microelectronics, for example, microelectromechanical systems (MEMs) could change our lives as long as we make use of all their possibilities. These include microsensors, back up memory cards and biomedical devices (defibrillators, neural simulators, drug vectors, etc.). Two-dimensional (2D) microbatteries have been able to meet the energy needs of these devices for many years. However, they are becoming progressively more energy demanding as they are carrying out an increasing number of functions. In addition, due to the increasing miniaturization of electronic devices, more energy must be produced on smaller surfaces. One solution to overcome this problem, just like the growth of skyscrapers in cities whose expansion surface is limited such as Hong Kong, is to move from a 2D approach to a three-dimensional (3D) approach [YAN 13]. In the hope of creating more efficient micro batteries or micro super capacitors, developing 3D architectures has become the main aim of the network, with the possibility of using some information acquired for 3D miniaturized systems in the design of thick electrodes for Li-ion batteries. Quite impressive research has already been obtained in this way (Figure 6.2). To do this, a current collector made of platinum has been deposited on these 3D structures by Atomic Layer Deposition (ALD). The advantage of these structures was then successfully assessed for two different microstorage devices. One contained a titanium dioxide electrode (TiO_2) formed by ALD for the 3D microbattery [BER 13] and another, the conformal electrolytical deposition of manganese dioxide (MnO_2) [BER 13].

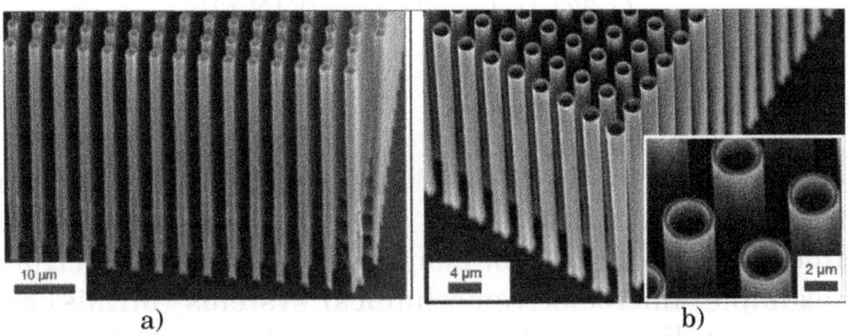

Figure 6.2. *Microstructures 3D machined in a Si substrate with micropillars a) and microtubes b) (source: Advanced Energy Materials)*

7

Technology Transfer, Research Promotion and Education

7.1. Development: industrial property

One of the reasons behind the creation of RS2E was the need to quickly transfer the results from academic research laboratories and technology transfer organizations to our Industrial partners; it was therefore necessary to define development politics for Intellectual Property (IP), agreed by all partners. This brought about the development unit, controlled by a patent engineer. The RS2E development structure is not independent but under the aegis of the CNRS. Its role is fundamental as it allows high responsiveness for filing patents, which can be exercised in one month. The RS2E development plan illustrated in the diagram below is as follows: a public research laboratory files a patent which is added to the RS2E patent portfolio. From the date it is introduced into the portfolio, it is offered only to Industrial Partners. Thus, each of RS2E's Industrial Partners knows the existence and contents of patent applications upon filing, providing a significant concurrential advantage over non-members due to the consistent 18 month confidentiality period after the filing of any application. Industrial Partners have the possibility of opting to use a patented technology during the first year of the patent's life (phase 1). If an option is taken, an R&D contract is created between the industrial and the inventors so that the industrial can, under restricted conditions, acquire a license on the patented technology (phase 2). At the end of phase 2 and the development contract, the industrial has the possibility of exercising an option by taking a license on the technology. If no option is taken (phase 1) or exercised (phase 2), the patent can be used beyond the circle of the Industrial Partners.

This industrial property management tool together with the existence of pre-transfer cells, which allow different stages in the design of a technology to be validated at any moment, is an innovative management tool designed for efficient and controlled industrialization.

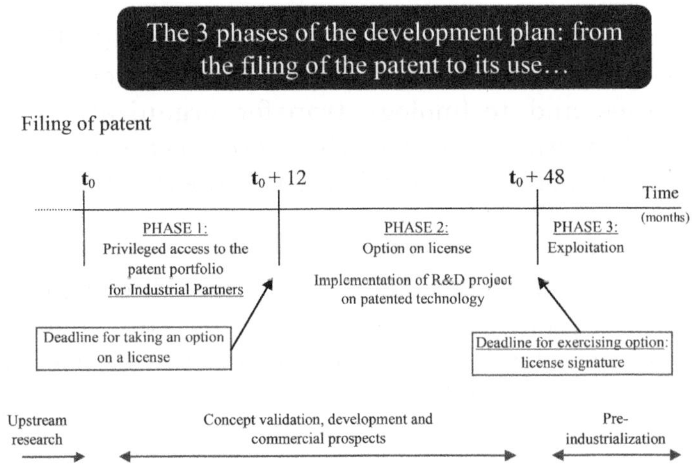

Figure 7.1. *Organization of intellectual property management: RS2E's Industrial partners benefit from prioritory information on filed patents compared to non-member industrials*

7.2. Education

Besides research and the transfer of technology, this network, home to proliferating scientific ideas and technological development, offers excellent training in the universities hosting the RS2E founding laboratories. The fundamental aspect of our training is already covered by Erasmus Mundus (MESC) whose scope includes "materials for energy storage and conversion". This activity was supported by the creation of RS2E through the implementation of block-release training in engineering. The aim of these two courses, the content of which we will describe next, is to prepare new generations of qualified scientists in relevant fundamental, analytical and technological aspects of energy storage, in which our current and future industries will need to be competitive.

7.2.1. *Erasmus Mundus Master's degree: Materials for Energy Storage and Conversion (MESC)*

The *Erasmus Mundus Master's degree: Materials for Energy Storage and Conversion* (MESC) was created in 2004 by the Alistore research network[1] (www.alistore.eu) and has since been coordinated by Professor Christian Masquelier of the University of Picardie Jules Verne (http://www.u-picardie.fr/mundus_MESC/). The MESC Master's degree is realized in conjunction with the universities of Toulouse III, Aix-Marseille, Amiens (France), and Warsaw (Poland) and Cordoba (Spain) in Europe.

The strong European and international component to the MESC Master's degree is largely responsible for its success, for more than one reason. The courses are conducted in English, which attracts excellent students from all over the world. The label, Erasmus Mundus, then provides funding and scholarships to selected students, which is obviously very attractive. Finally, students of this Master acquire a European culture, an open mind and scientific curiosity

1 The European Research Institute, Alistore-ERI, created in 2008, is a European Research Institute under the aegis of the CNRS (FR 3104) and includes 20 research laboratories and 14 European industries, carrying out research mainly on Li-ion batteries.

throughout the MESC course that leads to mobility between the universities of the consortium each semester. The final semester of the Master is a 6-month research internship that the students undertake in one of the academic laboratories or partner industries at RS2E and Alistore, or within the Universities of Drexel or Xiamen. This internship or "Master thesis" is fundamental because it is during this period that students have their first contact with the research environment.

Upon finishing the Master's degree, most students (74%) pursue a PhD, as shown in Figure 7.2. Even if they all do not stay within RS2E or Alistore research laboratories due to the many opportunities throughout Europe and the United States, the MESC Master's degree provides the chance to recruit high-class PhD students, with a strong background in solid-state chemistry and material science.

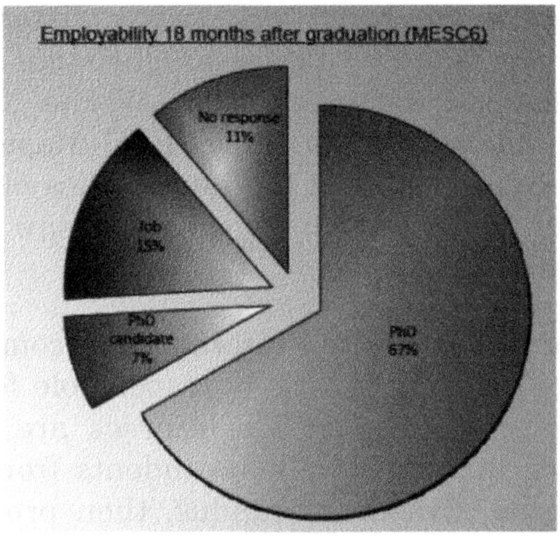

Figure 7.2. *Situation of MESC students 18 months after graduating. Most of them follow up with a PhD (6th class)*

The attractiveness of the MESC Master's degree can be measured through the diversity of students selected. European students are mainly from partner universities

(France, Poland and Spain), whereas non-European students come from all over the world, with Asia (China, India and Taiwan) in first place then South America (Mexico, Colombia and Brazil). Africa and Indonesia are also well represented. The wide variety of students provides our laboratories with the opportunity to recruit high-class students.

7.2.2. *Specialization in Energy Storage and Conversion (SCE), at ENSCBP (Bordeaux – INP)*

Specialization in Energy Storage and Conversion (SCE), at ENSCBP (Bordeaux – INP) is coordinated by L. Guerlou-Demourgues and L. Croguennec of the ICMCB laboratory. It is accessible to engineering students, at Master's level, at ENSCBP and 19 chemistry schools from *Fédération Gay-Lussac*, and also open to professionals in continuous training. Unlike the "research" specialization developed by the MESC Master's degree, this specialization aims to, together with the RS2E network and its objectives, train future engineers with a wider view of the problems associated with energy storage. Emphasis is first made on the assimilation of scientific and technical knowledge on different devices as well as materials. This training also allows students to acquire broad technical knowledge so that they are capable of solving problems associated with energy storage (such as the choice of materials or systems for a given storage application) and of professionally assessing it in the field.

The module involves 200 teaching hours, including many interventions by scientist from the private-sector. One Industry Project is undertaken by groups of students working on a current topic proposed by an industrial. The subject may be bibliographic and/or on the latest technological breakthroughs. The research undertaken must use the project management approach used in the private-sector. It is vocational and may eventually lead to an internship in a sponsor company.

Conclusion

Energy transition is a reality and all countries are in a position to effectively address it because it brings about the start of a green economy full of jobs and hope of economic recovery. Success in this task will require synergizing all links in the chain to go beyond the production of energy, to its management, use and storage. We have only discussed the national initiative taken by the French Ministry of Higher Education and Research and CNRS in the creation of RS2E to gather and unite French efforts in energy storage to innovate and make our country a leader of tomorrow. Much hope now rests on the electrochemical storage, which appears as the best choice for electric vehicles and, perhaps, for grid applications, clearly suggesting very lucrative business.

This new scientific organization in the field of energy storage has changed the way we work and requires researchers to be not only excellent but also to promote interdisciplinarity, mobility, sharing of equipment and students, space and even intellectual property. A dramatic change is, therefore, possible. By sharing all of this, it will lead to the strong integration of our country's scientific resources toward a common goal, and not just any goal: to better manage the energy resources of our planet. These are issues that are more than ever at the center of our daily concerns and represent a major socioeconomic priority.

Three years after its launch, it would be premature to declare victory, but we hope to have convinced our readers through our way of working, the variety and quality of our research, that we are on the right track. This partnership enables us to advance in topics that we could not have addressed previously (1) due to a too low critical mass, (2) lack of combined expertise and (3) the lack of exchange and communication of information, a key component in innovation. The establishment of strong and numerous interactions with industry and technology transfer centers via transfer units and platforms now allows the rapid transition from concept to product and thus the ability to be competitive. The Na-ion project is a typical example of what would not have been possible without RS2E.

Our biggest challenge is that we will race against the clock because often the deadline given to us by our institutions is 2050. What chance do we have? Fortunately, we have a rich periodic table of elements. Admittedly, this is a great opportunity as well as a nightmare to find the winning composition, given the large number of possible combinations. This is why RS2E rests much hope on rational combinatorial theoretical chemistry to establish a genome of materials.

Despite the achievements that can be expected of a predictive theoretical approach, it is still up to the solid-state chemist to design clever synthesis pathways for development of new materials as well as to understand the mechanisms, which is why RS2E has developed advanced analytical platforms. Theory-experiment-observation certainly form an ideal triad, satisfying our intellect and allowing new compounds to be prepared, but it is unproductive if useful materials are not actually used. Therefore, systemic approaches involving the cooperation and multiple disciplines are required, hence the importance of pretransfer units and the evolving nature of the network, which is always ready to integrate new experts.

Finally, note our leadership in this national organization through the creation of RS2E, since major projects have been set up in Europe (in particular, Germany), United States with the creation of JCESR (Joint Center for Energy Storage Research) as well as in Japan with the RISING network (Research & Development Initiative for Scientific Innovation of New Generation Battery). Huge investments were made in order to better address this new gold rush, the green economy. We should be delighted to have been able and known how to organize ourselves quickly to address this global competition. Now, it is up to us to use this advance to build an efficient, sustainable and prosperous network, as Henry Ford once said: "Coming together is the beginning. Keeping together is progress. Working together is success, but building together is prosperity". Innovation for prosperity, such is our motto.

Bibliography

[ABO 11] ABOULAICH A., BOUCHET R., DELAIZIR G., *et al.*, "A new approach to develop safe all-inorganic monolithic Li-ion batteries", *Advanced Energy Materials*, vol. 1, pp. 179–183, 2011.

[ANA 87] ANANI A., CROUCH-BAKER S., HUGGINS R.A., "Kinetics and thermodynamic parameters of several binary alloys negative electrode materials at ambient temperature", *Journal of the Electrochemical Society*, vol. 134, pp. 3098–3102, 1987.

[ARM 79] ARMAND M.B., CHABAGNO J.M., DUCLOT M., "Poly-ethers as solid electrolytes", in VASHITSHTA P., MUNDY J.N., SHENOY G.K., *Fast Ion Transport in Solids, Electrodes and Electrolytes*, North Holland Publishers, Amsterdam, 1979.

[ARM 08] ARMAND M., TARASCON J.-M., "Building better batteries", *Nature*, vol. 451, pp. 652–657, 2008.

[ARM 09] ARMAND M., GRUGEON S., VEZIN H., *et al.*, "Conjugated dicarboxylate anodes for Li-ion batteries", *Nature Materials*, vol. 8, no. 2, pp. 120–125, 2009.

[AUG 13] AUGUSTYN V., COME J., LOW M.A., *et al.*, "Evidence of intercalation pseudocapacitance in high-rate lithium-ion energy storage materials", *Nature Materials*, vol. 6, pp. 518–522, 2013.

[BAR 89] BARTOLOZZI M., "Development of redox flow batteries – a historical bibliography", *Journal of Power Sources*, vol. 27, p. 219, 1989.

[BAR 11] BARPANDA P., ATI M., MELOT B.C., *et al.*, "A 3.90 V iron-based fluorosulphate material for lithium-ion batteries crystallizing in the triplite structure", *Nature Materials,* vol. 10, no. 10, pp. 772–779, 2011.

[BER 13] BERNARDI M., PALUMMO M., GROSSMAN J.C., "extraordinary sunlight absorption and one nanometer thick photovoltaics using two-dimensional monolayer materials", *Nano Letters*, vol. 13, pp. 3664–3670, 2013.

[BET 84] BETZ S., TRIBUTSH H., FIECHTER S., "Photo-intercalation and optical information storage using Cu6-xPS5I", *Journal of The Electrochemical Society: Solid State Science and Technology – Technical Papers*, vol. 3, no. 131, pp. 640–644, 1984.

[BON 13] BONNARD G., BARRÈS A.-L., MENTRÉ O., *et al.*, "The low/room-temperature forms of the lithiated salt of 3,6-dihydroxy-2,5-dimethoxy-p-benzoquinone: a combined experimental and dispersion-corrected density functional study", *CrystEngComm*, vol. 15, pp. 2809–2821, 2013.

[BOU 13] BOULINEAU S., TARASCON J.-M., LERICHE J.-B., *et al.*, "Electrochemical properties of all-solid-state lithium secondary batteries using Li-argyrodite Li6PS5Cl as solid electrolyte", *Solid State Ionics*, vol. 242, pp. 45–48, 2013.

[BRO 10] BROUSSE T., "Pseudocapacitive materials for supercapacitor applications", *International Conference on Advanced Capacitors (ICAC2010), Meeting abstract*, Kyoto, Japan, 31 May–2 June 2010.

[BRU 12] BRUCE P.G., FREUNBERGER S.A., HARDWICK L.J., *et al.*, "Li-O_2 and Li-S batteries with high energy storage", *Nature Materials*, vol. 11, no. 1, pp. 19–29, 2012.

[CHE 08] CHEN H., ARMAND M., DEMAILLY G., *et al.*, "From biomass to a renewable $Li_xC_6O_6$ organic electrode for sustainable Li-ion batteries", *ChemSus Chem*, vol. 1, no. 4, pp. 348–355, 2008.

[CHM 06] CHMIOLA J., YUSHIN G., GOGOTSI Y., *et al.*, "Anomalous increase in carbon capacitance at pore size below 1 nm", *Science*, vol. 313, pp. 1760–1763, 2006.

[COM 14] COME J., AUGUSTYN V., KIM J.W., *et al.*, "Electrochemical kinetics of nanostructured Nb_2O_5 electrodes", *Journal of the Electrochemical Society*, vol. 161, pp. A718–A725, 2014.

[DAH 91] DAHN J.R., VON SACKEN U., JUZKOW M.W., *et al.*, "Rechargeable $LiNiO_2$/carbon cells", *Journal of the Electrochemical Society,* vol. 138, pp. 2207–2211, 1991.

[DAR 12] DARWICHE A., MARINO C., SOUGRATI M.T., *et al.*, "Better cycling performances of bulk Sb in Na-ion batteries compared to Li-ion systems: an unexpected electrochemical mechanism", *Journal of American Chemical Society*, vol. 134, pp. 20805–20811, 2012.

[DEL 80] DELMAS C., FOUASSIER C., HAGENMULLER P., "Structural classification and properties of the layered oxides", *Physica B & C*, vol. 99, nos. 1–4, pp. 81–85, 1980.

[DEM 11] DEMIR-CAKAN R., MORCRETTE M., NOUAR F., *et al.*, "Cathode composites for Li-S batteries via the use of oxygenated porous architectures", *Journal of the American Chemical Society*, vol. 133, pp. 16154–16160, 2011.

[DEM 13] DEMARCONNAY L., CALVO E.G., TIMPERMAN L., *et al.*, "Optimizing the performance of supercapacitors based on carbon electrodes and protic ionic liquids as electrolytes", *Electrochimica Acta,* vol. 108, pp. 361–368, 2013.

[DEM 14] DEMIR-CAKAN R., MORCRETTE M., LERICHE J.B., *et al.*, "An aqueous electrolyte rechargeable Li-ion/polysulfide battery", *Journal of Materials Chemistry A*, vol. 2, no. 24, pp. 9025–9029, 2014.

[DUD 11] DUDUTA M., HO B., WOOD V.C., *et al.*, "Semi-solid lithium rechargeable flow battery", *Advanced Energy Materials*, vol. 1, p. 511, 2011.

[DUN 11] DUNN B., KAMATH H., TARASCON J.-M., "Electrical energy storage for the grid: a battery of choices", *Science*, vol. 334, no. 6058, p. 928, 2011.

[EUS 14] EUSTACHE E., TILMANT P., MORGENROTH L., "Silicon-microtube scaffold decorated with anatase TiO_2 as a negative electrode for a 3d litium-ion microbattery", *Advanced Energy Materials*, vol. 4, pp. 1301612–130162, 2014.

[GAO 14] GAO P.-C., RUSSO P.A., CONTE D.E., *et al.*, "Morphology effects on the supercapacitive electrochemical performances of iron oxide/reduced graphene oxide nanocomposites", *ChemElectroChem*, vol. 1, no. 4, pp. 747–754, 2014.

[GHI 11] GHIMBEU C.M., RAYMUNDO-PINERO E., FIOUX P., *et al.*, "Vanadium nitride/carbon nanotube nanocomposites as electrodes for supercapacitors", *Journal of Materials Chemistry,* vol. 21, pp. 13268–13275, 2011.

[GHI 12] GHIMBEU C.M., SIMA F., OSTACI R.V., *et al.*, "Crystalline vanadium nitride ultra-thin films obtained at room temperature by pulsed laser deposition", *Surface and Coatings technology,* vol. 211, pp. 158–162, 2012.

[GHI 14] GHIMBEU C.M., MALAK-POLACZYK A., FRACKOWIAK E., *et al.*, "Template-derived high surface area lambda-MnO_2 for supercapacitor applications", *Journal of Applied Electrochemistry*, vol. 44, pp. 123–132, 2014.

[GHO 12] GHODBANE O., ATAHÉRIAN F., WU N.L., *et al.*, "In situ crystallographic investigations of charge storage mechanisms in MnO2-based electrochemical capacitors", *Journal of Power Sources*, vol. 206, pp. 545–462, 2012.

[GOD 11] GODILLOT J., DELMAS C., SIMON P., *et al.*, "Original conductive nano-Co_3O_4 investigated as electrode material for hybrid supercapacitors", *Electrochemical and Solid State Letters*, vol. 14, pp. A139–A142, 2011.

[HAM 12] HAMELET S., TZEDAKIS T., LERICHE J.-B. *et al.*, "Non-aqueous Li-based redox flow batteries", *Journal of The Electrochemical Society*, vol. 159, no. 8, pp. A1360–A1367, 2012.

[HUA 09] HUANG H., FAULKNER T., BARKER J. *et al.*, "Lithium metal phosphates, power and automotive applications", *Journal of Power Sources*, vol. 189, p. 748, 2009.

[HUI 12] HUI W., CUI Y., "Designing nanostruchered Si anodes for high energy lithium ion batteries", *Nanotoday,* vol. 7, no. 5, pp. 414–429, 2012.

[JI 10] JI X., NAZAR L.F., "Advances in Li-S batteries", *Journal of Materials Chemistry*, vol. 20, pp. 9821–9826, 2010.

[JI 11] JI X., EVERS S., BLACK R., *et al.*, "Stabilizing lithium–sulphur cathodes using polysulphide reservoirs", *Nature Commun,* vol. 2, p. 325, 2011.

[JOH 14] JOHNSON L., LI C., LIU Z., *et al.*, "The role of LiO_2 solubility in O_2 reduction in aprotic solvents and its consequences for Li-O_2 batteries", *Nature Chemistry,* vol. 6, pp. 1091–1099, 2014.

[JME 09] JM ENERGY CORPORATION, Front runner of lithium ion capacitor, available at http://www.jmenergy.co.jp/en/, 2009.

[KAM 11] KAMAYA N., HOMMA K., YAMAKAWA Y., *et al.*, "A lithium superionic conductor", *Nature Materials*, vol. 10, pp. 682–686, 2011.

[KOM 11] KOMABA S., MURATA W., ISHIKAWA T., *et al.*, "Electrochemical Na insertion and solid electrolyte interphase for hard-carbon electrodes and application to Na-ion batteries", *Advanced Functional Materials*, vol. 21, no. 20, pp. 3859–3867, 2011.

[LAL 14] LALÈRE F., LERICHE J.B., COURTY M., *et al.*, "An all-solid state NASICON sodium battery operating at 200°C", *Journal of Power Sources*, vol. 247, pp. 975–980, 2014.

[LAR 15] LARCHER D., TARASCON J.M., "Towards greener and more sustainable batteries for electrical energy storage", *Nature Chemistry*, vol. 7, pp. 19–29, 2015.

[LEB 14] LEBÈGUE E., BROUSSE T., GAUBICHER J., *et al.*, "Toward fully organic rechargeable charge storage devices based on carbon electrodes grafted with redox molecules", *Journal of Materials Chemistry A*, vol. 2, pp. 8599–8602, 2014.

[LEG 01] LEGAGNEUR V., AN Y., MOSBAH A., *et al.*, "LiMBO3 (M = Mn, Fe, Co): synthesis, crystal structure and lithium deinsertion/insertion properties", *Solid State Ionics,* vol. 139, nos. 1–2, pp. 37–46, 2001.

[LIN 11] LIN R., TABERNA P.-L., FANTINI S., *et al.*, "Capacitive energy storage from -50°C to 100°C using an ionic liquid electrolyte", *Journal of Physical Chemistry Letters*, vol. 19, pp. 2396–2401, 2011.

[LIU 14] LIU N., ZHENDA L., ZHAO J., *et al.*, "A pomegranate-inspired nanoscale design for large-volume-change lithium battery anodes", *Nature Nanotechnology*, vol. 9, pp. 187–192, 2014.

[LON 04] LONG J.W., DUNN B., ROLISON D.R., *et al.*, "Three-dimensional battery architectures", *Chemical Review*, vol. 104, p. 4463, 2004.

[LUC 12] LUCIO PORTO R., BOUHTIYYA S., PIERSON PJ.F., *et al.*, "VN thin films as electrode materials for electrochemical capacitors", *Electrochimica Acta,* vol. 82, pp. 257–262, 2012.

[MA 93] MA Y., DOEFF M.M., VISCO S.J., *et al.*, "Electrochemical insertion of sodium into carbon", *Journal of Electrochemical Society*, vol. 140, pp. 2726–2733, 1993.

[MAR 13] MARINO C., SOUGRATI M.T., DARWICHE A., *et al.*, "Study of the series Ti1-yNbySnSb with 0 <= y <= 1 as anode material for Li-ion batteries", *Journal of Power Sources,* vol. 244, pp. 736–741, 2013.

[MAS 13] MASQUELIER C., CROGUENNEC L., "Polyanionic (phosphates, silicates, sulfates) frameworks as electrode materials for rechargeable Li (or Na) batteries", *Chemical Reviews*, vol. 113, pp. 6552–6591, 2013.

[MER 12] MERLET C., ROTENBERG B., MADDEN P.A., *et al.*, "On the molecular origin of supercapacitance in nanoporous carbon electrodes", *Nature Materials,* vol. 11, pp. 306–310, 2012.

[MER 13] MERLET C., PEAN C., ROTENBERG B., *et al.*, "Hidden ions store charge more efficiently in supercapacitors", *Nature Communications,* vol. 4, p. 2701, 29 October 2013.

[MIK 08] MIKHAYLIC Y.V., Electrolytes for lithium sulfur cells, US patent no. 7354680, 2008.

[MIO 14] MIOT J., RECHAM N., LARCHER D., *et al.*, "Biomineralized a-Fe_2O_3: texturation and electrochemical reaction with Li", *Energy and Environmental Science*, vol. 7, pp. 451–9943, 2014.

[MIZ 80] MIZUSHIMA K., JONES P.C., WISEMAN P.J., *et al.*, "LixCoO2 (0<x 1): a new cathode for batteries of high energy density", *Materials Research Bulletin,* vol. 15, pp. 783–789, 1980.

[MUR 78] MURPHY D.W., DISALVO F.J., CARIDES J.N., *et al.*, "Topochemical reactions of rutile related structures with lithium", *Materials Research Bulletin,* vol. 13, pp. 1395–1402, 1978.

[NAG 90] NAGAURA T., TOZAWA K., "Lithium ion rechargeable battery", *Progress in Batteries and Solar Cells,* vol. 9, p. 209, 1990.

[NAM 06] NAM K.T., KIM D.-W., YOO P.J., *et al.*, "Virus-enabled synthesis and assembly of nanowires for lithium ion battery electrodes", *Science*, vol. 312, p. 885, 2006.

[NGU 13] NGUYEN B.P.N., KUMAR N.A., GAUBICHER J., *et al.*, "Nanosilicon-based thick negative composite electrodes for lithium batteries with graphene as conductive additive", *Advanced Energy Materials,* vol. 3, pp. 1351–1357, 2013.

[NOM 95] NOMIYAMA T., KURIYAKI H., HIRAKAWA K., "Photorechargeable battery using new layer compound $CuFeTe_2$", *Conference: International Conference on Science and Technology of Synthetic Metals (ICSM 94), Synthetic Metals*, vol. 71, nos. 1–3, pp. 2237–2238, 1995.

[NYT 05] NYTEN A., ABOUIMRANE A., ARMAND M., *et al.*, "Electrochemical performance of Li_2FeSiO_4 as a new Li-battery cathode material", *Electrochemistry Communication*, vol. 7, no. 2, pp. 156–160, 2005.

[OGA 06] OGASAWARA T., DEBART A., HOLZAPFEL M., *et al.*, "Rechargeable Li_2O_2 electrode for lithium batteries", *Journal of the American Chemical Society*, vol. 128, no. 4, pp. 1390–1393, 2006.

[OHZ 01] OHZUKU T., MAKIMURA Y., YAMAMOTO S., *et al.*, "A 3-volt lithium-ion cell with $Li[Ni_{1/2}Mn_{3/2}]O_4$ and $Li[Li_{1/3}Ti_{5/3}]O_4$: a method to prepare stable positive-electrode material of highly crystallized $Li[Ni_{1/2}Mn_{3/2}]O_4$", *Chemistry Letters*, vol. 30, no. 12, pp. 1270–1271, 2001.

[PAN 14] PANG Q., KUNDU D., CUISINIER M., *et al.*, "Surface-enhanced redox chemistry of polysulphides on a metallic and polar host for lithium-sulphur batteries", *Nature Communications*, vol. 5, p. 4759, 2014.

[PAT 13] PATRA S., DAVOISNE C., BRUYERE S., *et al.*, "Room-temperature synthesis of high surface area anatase TiO_2 exhibiting a complete lithium insertion solid solution", *Particle and Particle Sytems Characterization*, vol. 30, no. 12, pp. 1093–1104, 2013.

[PEN 12] PENG Z., FREUNBERGER S.A., CHEN Y., *et al.*, "A reversible and higher-rate Li-O_2 battery", *Science,* vol. 337, no. 6094, pp. 563–566, 2012.

[PÉT 12] PÉTRISSANS X., BÉTARD A., GIAUME D., *et al.*, "Solution synthesis of nanometric layered cobalt oxides for electrochemical applications", *Electrochimica Acta,* vol. 66, pp. 306–312, 2012.

[POG 12] POGNON G., COUGNON C., MAYILUKILA D., *et al.*, "Catechol-modified activated carbon prepared by the diazonium chemistry for application as active electrode material in electrochemical capacitor", *ACS Applied Materials & Interfaces,* vol. 4, pp. 3788–3796, 2012.

[POI 01] POIZOT P., LARUELLE S., GRUGEON S., *et al.*, "Searching for new anode materials for Li-ion technology: time to deviate from the usual path", *Journal of Power Sources,* vol. 97, pp. 235–239, 2001.

[POI 11] POIZOT P., DOLHEM F., "Clean energy new deal for a sustainable world: from non-CO_2 generating energy sources to greener electrochemical storage devices", *Energy and Environmental Science,* vol. 4, p. 2003, 2011.

[PON 12] PONROUCH A., MARCHANTE E., COURTY M., *et al.*, "In search of an optimized electrolyte for Na-ion batteries", *Energy and Environmental Science*, vol. 5, pp. 8572–8583, 2012.

[PON 13a] PONROUCH A., DEDRYVÈRE R., MONTI D., *et al.*, "Towards high energy density sodium ion batteries through electrolyte optimization", *Energy and Environmental Science,* vol. 6, p. 2361, 2013.

[PON 13b] PONROUCH A., GOÑI A.R., PALACÍN M.R., "High capacity hard carbon anodes for sodium ion batteries in additive free electrolyte", *Electrochemistry Communications*, vol. 27, pp. 85–88, 2013.

[RAV 99] RAVET N., GOODENOUGH J.B., BESNER S., *et al.*, "Improved iron-based cathode material", p. 127, *ECS Fall Meeting Abstract*, Hawaii, 1999.

[REC 10] RECHAM N., CHOTARD J.N., DUPONT L., *et al.*, "A 3.6 V lithium-based fluorosulphate insertion positive electrode for lithium-ion batteries", *Nature Materials*, vol. 9, pp. 68–74, 2010.

[RED 09] REDDY M.A., PRALONG V., CAIGNAERT V., *et al.*, "Monoclinic iron hydroxy sulphate: a new route to electrode materials", *Electrochemistry Communications*, vol. 11, no. 9, pp. 1807–1810, 2009.

[REN 13] RENAULT S., GOTTIS S., BARRÈS A.-L., *et al.*, "A green Li-organic battery working as a fuel cell in case of emergency", *Energy and Environmental Science*, vol. 6, pp. 2124–2133, 2013.

[ROU 14] ROUSSE G., TARASCON J.M., "Sulfate-based polyanionic compounds for Li-ion batteries: synthesis, crystal chemistry, and electrochemistry aspects", *Chemistry of Materials*, vol. 26, pp. 394–406, 2014.

[SAF 15] SAFT, SNC–Ahead with supercapacitors, available at http://www.saftbatteries.se/battery-search/snc, 2015.

[SAT 13a] SATHIYA M., RAMESHKA K., ROUSSE G., *et al.*, "High performance $Li_2Ru_{1-y}Mn_yO_3$ (0.2 <= y <= 0.8) Cathode materials for rechargeable lithium-ion batteries: their understanding", *Chemistry of Materials*, vol. 25, pp. 1121–1131, 2013.

[SAT 13b] SATHIYA M., ROUSSE G., RAMESHKA K., *et al.*, "Reversible anionic redox chemistry in high-capacity layered-oxide electrodes", *Nature Materials*, vol. 12, pp. 827–835, 2013.

[SAT 14] SATHIYA M., ABAKUMOV A.M., FOIX D., *et al.*, "Origin of voltage decay in high-capacity layered oxide electrodes", *Nature Materials*, December 2014 (advanced online publication).

[SEN 11] SENGUTTUVAN P., ROUSSE G., SEZNEC V., *et al.*, "$Na_2Ti_3O_7$: lowest voltage ever reported oxide insertion electrode for sodium ion batteries", *Chemistry of Materials*, vol. 23, pp. 4109–4111, 2011.

[SIM 08] SIMON P., GOGOTSI Y., "Materials for electrochemical capacitors: building a battery of the future", *Nature Materials*, vol. 7, pp. 845–854, 2008.

[SIM 13] SIMON P., GOGOTSI Y., "Capacitive energy storage in nanostructured carbon-electrolyte systems", *Account for Chemical Research*, vol. 5, pp. 1094–1103, 2013.

[SKY 86] SKYLLAS-KAZACOS M., RYCHCIK M., ROBINS R.G., *et al.*, "New all-vanadium redox flow cell", *Journal of the Electrochemical Society*, vol. 133, p. 1057, 1986.

[TAO 14] TAO X.Y., WANG J., YING Z., *et al.*, "strong sulfur binding with conducting magneli-phase Ti_nO_{2n-1} nanomaterials for improving lithium-sulfur batteries", *Nano Letters*, vol. 14, no. 9, pp. 5288–5294, 2014.

[TAR 96] TARASCON J.-M., GOZDZ A.S., SCHMUTZ C., *et al.*, "Performance of Bellcore's plastic rechargeable Li-ion batteries", *Solid State Ionics,* vol. 86–88, pp. 49–54, 1996.

[TAR 98] TARASCON J.-M., PERCHERON-GUEGAN A., "Electrode materials for performing rechargeable batteries, a scientific and industrial challenge", *Actualité Chimique*, vol. 3, pp. 12–14, 1998.

[TAR 01] TARASCON J.-M., ARMAND M., "Issues and challenges facing rechargeable lithium batteries", *Nature*, vol. 414, pp. 359–367, 2001.

[TAR 08] TARASCON J.-M., "Towards sustainable and renewable systems for electrochemical energy storage", *ChemSusChem*, vol. 1, pp. 777–779, 2008.

[TAR 10] TARASCON J.-M., RECHAM N., ARMAND M., *et al.*, "Hunting for better Li-based electrode materials via low temperature inorganic synthesis", *Chemistry of Materials*, vol. 22, pp. 724–739, 2010.

[TAR 11] TARASCON J.-M., Leçon Inaugurale, Collège de France, 2011.

[THA 83] THACKERAY M.M., DAVID W.I.F., BRUCE P.G., *et al.*, "Lithium insertion into manganese spinels", *Materials Research Bulletin,* vol. 18, pp. 461–472, 1983.

[THA 05] THACKERAY M.M., JOHNSON C.S., VAUGHEY J.T., *et al.*, "Advances in manganese-oxide 'composite' electrodes for lithium-ion batteries", *Journal of Materials Chemistry,* vol. 15, p. 2257, 2005.

[THO 13] THOTIYL M.M.O., FREUNBERGER S.A., PENG Z., *et al.*, "A stable cathode for the aprotic Li-O_2 battery", *Nature Materials,* 1 September 2013.

[TOU 15] TOUIDJINE A., LARCHER D., *et al.*, "Partially oxidized silicon particles for stable water slurries and practical large-scale making of Si/CMC/C electrodes", 2015.

[TSA 13a] TSAI W.-Y., GAO P.-C., DAFFOS B., *et al.*, "Ordered mesoporous silicon carbide-derived carbon for high-power supercapacitors", *Electrochemistry Communications,* vol. 34, pp. 109–112, 2013.

[TSA 13b] TSAI W.-Y., LIN R., MURALI S., *et al.*, "Outstanding performance of activated graphene based supercapacitors in ionic liquid electrolyte from –50 to 80°C", *Nano Energy,* vol. 2, pp. 403–411, 2013.

[TSA 14] TSAI W.-Y., TABERNA P.L., SIMON P., *et al.*, "Electrochemical quartz crystal microbalance (EQCM) study of ion dynamics in nanoporous carbons", *Journal of the American Chemical Society*, vol. 136, pp. 8722–8728, 2014.

[UDD 13] UDDIN J., BRYANTSEV V.S., GIORDANI V., *et al.*, "Lithium nitrate as regenerable SEI stabilizing agent for rechargeable Li/O_2 batteries", *Journal of Physical Chemistry and Letters,* vol. 4, pp. 3760–3765, 2013.

[VIS 13] VISCO S.J., NIMON Y.S., KATZ B.D., *et al.*, Aqueous electrolyte lithium sulfur batteries, US Patent no. 2013/ 0122334 A1, 2013.

[WEI 12] WEISSMANN M., CROSNIER O., BROUSSE T., *et al.*, "Electrochemical study of anthraquinone groups, grafted by the diazonium chemistry, in different aqueous media-relevance for the development of aqueous hybrid electrochemical capacitor", *Electrochimica Acta*, vol. 82, pp. 250–256, 2012.

[WIN 99] WINTER M., BESENHARD J.O., "Electrochemical lithiation of tin and tin-based intermetallics and composites", *Electrochimica Acta,* vol. 45, pp. 31–50, 1999.

[YAB 12] YABUUCHI N., KAJIYAMA M., IWATATE J., *et al.*, "P2-type $Na_x[Fe_{1/2}Mn_{1/2}]O_2$ made from earth-abundant elements for rechargeable Na batteries", *Nature Materials*, vol. 11, no. 6, pp. 512–517, 2012.

[YAN 13] YAN N.F., LI G.R., GAO X.P., "Solar rechargeable redox flow battery based on Li_2WO_4/LiI couples in dual-phase electrolytes", *Journal of Materials Chemistry A*, vol. 1, p. 7012, 2013.

[YAO 67] YAO Y.F.Y., KUMMER J.T., "Ion exchange properties of and rates of ionic diffusion in beta-alumina", *Journal of Inorganic & Nuclear Chemistry*, vol. 29, no. 9, p. 2453, 1967.

[YOU 13] YOUSSRY M., MADEC L., CERBELAUD M., *et al.*, "Non-aqueous carbon black suspensions for lithium-based redox flow batteries: rheology and simultaneous rheo-electrical behavior", *Physical Chemistry Chemical Physics*, vol. 15, pp. 14476–14486, 2013.

Index

B, C, D, E

bioinspired approach, 41, 42–45
CEA, 32
CNRS, 34, 53, 55, 59
dendrite, 2
electrolyte for batteries, 1–4, 8–10
electrolyte for supercapacitors, 20, 21
energy density of supercapacitors, 20, 22

I, L, M

interfaces, 15, 32, 37–40, 47
ionic conductors, 4, 37, 39
ionic liquids, 21, 42, 44
lamellar compounds, 11, 13, 30, 31
Li-air batteries, 25
Li-ion batteries, 24, 34, 37, 45–47, 52, 55
Li-metal batteries, 2, 50
Li-S batteries, 29
$LiCoO_2$ positive electrode, 34
$LiFePO_4$ positive electrode, 7
modeling, 17, 22

N, O, P

Na-ion batteries, 30, 32, 38
nanometric (materials for batteries), 6, 7, 11, 26, 43, 46
NMC, 5, 13, 14
organic (materials for batteries), 2, 10, 12, 18–22, 37, 43, 45, 46
Pb-acid batteries, 1
porosity of carbon and supercapacitors, 8, 17, 20
pseudocapacitive materials, 9, 17–20, 22
pumped-storage, 23, 32, 33

R, S

redox-flow batteries, 33
solid state batteries, 36–40

Spark Plasma Sintering (SPS), 37
stationary storage (grid storage), 4
supercapacitors, 1, 17, 20–22
sustainable development, 12, 23, 41, 42, 45

Printed and bound by CPI Group (UK) Ltd, Croydon, CR0 4YY

13/09/2026

14961080-0001